Electronics Construction and Assembly

Electronics Construction and Assembly

GEORGE L. RITCHIE

San Jose City College

PRENTICE-HALL, INC., Englewood Cliffs, New Jersey 07632

Library of Congress Cataloging in Publication Data

Ritchie, George L
Electronics construction and assembly.

1. Electronic apparatus and appliances—Design and construction. I. Title.
TK7870.R55 1980 621.381 79-12594
ISBN 0-13-250472-3

Editorial/production supervision by Virginia Rubens
Manufacturing buyer: Gordon Osbourne

Printed in the United States of America

10 9 8 7 6 5 4 3 2

PRENTICE-HALL INTERNATIONAL, INC., *London*
PRENTICE-HALL OF AUSTRALIA PTY. LIMITED, *Sydney*
PRENTICE-HALL OF CANADA, LTD., *Toronto*
PRENTICE-HALL OF INDIA PRIVATE LIMITED, *New Delhi*
PRENTICE-HALL OF JAPAN, INC., *Tokyo*
PRENTICE-HALL OF SOUTHEAST ASIA PTE. LTD., *Singapore*
WHITEHALL BOOKS LIMITED, *Wellington, New Zealand*

Contents

6 FINISHES *153*

7 LABELING AND MARKING *162*

PART FOUR PROJECTS

8 CONSTRUCTION AND ASSEMBLY OF PROJECTS *177*

PART FIVE BASIC SHOP TOOLS

9 THREAD-GAUGE FABRICATION *201*

10 TOOLS AND TOOL SHARPENING *220*

PART SIX APPENDIXES

A PRINTED-CIRCUIT INFORMATION *235*

B SUPPLEMENTARY INFORMATION *254*

C CHARTS AND TABLES *277*

Preface

This book contains detailed information necessary for instruction in electronics assembly and for the development of a prototype electronics unit. Mechanical aspects of assembly and construction are emphasized, and topics vital to a junior college or technical institute program have been carefully selected and presented in a practical and straightforward manner. The book covers the following areas:

Basic information on drafting.

Soldering and assembly techniques.

Printed circuits—artwork and fabrication.

Chassis construction and project or prototype work.

Chemical processes for finishing and etching.

Methods of labeling and marking.

Use of photo-sensitive emulsions.

Basic hand- and machine-tool operations.

Information sheets.

Short-answer student response sheets.

Introductory information and directions are presented with each assignment. The assignments deal with the problems encountered in different phases of fabrication and are designed to meet the informational and educational needs of the beginning electronics technician. It is expected that the student, or the instructor directing the student, will decide which assignments are most appropriate for his or her particular situation. The book can be used directly as a lab text by an instructor, as a reference, as a guide for a single operation, or to provide information relating to the fabrication of a complete chassis.

There is more than enough material to provide assignments for two semesters in a class consisting of one lecture and three laboratory hours per week. Included in the appendixes are a large number of useful and important information sheets.

The organized approach to the problems of prototype construction, the numerous examples, and the general and specific information included make this book readily adaptable to the present-day technical education programs in the electronics field. The author has used the assignments as the framework for a separate assembly and chassis construction class. However, the book is also designed to be used together with, and as a supplement to, an introductory electronics theory class. In any case, its versatility will prove valuable to the instructor.

Instructors in technical education programs should be turning out competent men and women, not simply producing electronics chassis or projects. Students should be taught to think, to plan, and to work out solutions. It would be wrong to allow students to get in the habit of mechanically following a step-by-step sequence list in their work. For this reason the assignments in this book have been carefully developed. The important, essential, and technical detail is given in each assignment, but in many of the assignments certain details have been left for the student to determine. Mistakes will probably be made, but this gives the instructor the chance to teach the students to think and it gives students a chance to learn. This leaves room for creative thinking, planning, and problem-solving and will prevent the development of a prototype, "cookbook" technician.

It is recommended that the work with drafting be started at the beginning of the semester. This gives the instructor an opportunity to introduce a useful and valuable topic without the need for any materials or equipment except a piece of $8\frac{1}{2} \times 11''$ plain paper, a No. 2 pencil, and a ruler. As time progresses in the first or second week, students should be getting together their own tools to work with. Subsequently, the assignments dealing with assembly, more advanced chassis work, printed circuit work, use of tools, and various finishes can be chosen to fit into any course arrangement. The assignments in the book are not necessarily set up on a sequential basis, and topics can be selected at the discretion of the instructor.

Those assignments marked (1) are basic and have been prepared for first-semester classes. Those assignments marked (2) are more advanced and are for second-semester classes. The unmarked assignments can be used in either situation.

Recognition should be given to Margaret Shumate, who gave invaluable help in proofreading the material for the book. Also, with gratitude for an excellent job, the author wishes to acknowledge the assistance of Marian Essex and Joy Nabi, typists, and Don Strong, draftsman.

Palo Alto, California GEORGE L. RITCHIE

PART ONE

INTRODUCTION TO ASSEMBLY

1

Drafting Techniques

BASIC ELEMENTS OF DRAFTING (1)*

The use of drawings is a means of communication. In many ways it is a more exacting and simpler way of communicating an idea than by use of the written or spoken word. As we all know, this means of communication has been in use since earliest times. In today's highly technical world we find that the use of drawings as a means of communication is vital, and even more important than the written or spoken word.

On drawing paper we can convey many ideas, such as *shape, size,* and *general construction,* by following a few commonly accepted standards or practices. Today's technicians need to know how to "use" drawings to portray ideas and also need to know how to "read" a drawing.

*Note: number following section heading indicates level of assignment: those marked (1) are basic and are for first-semester classes; those marked (2) are more advanced and are for second-semester classes. Unmarked assignments can be used in either situation.

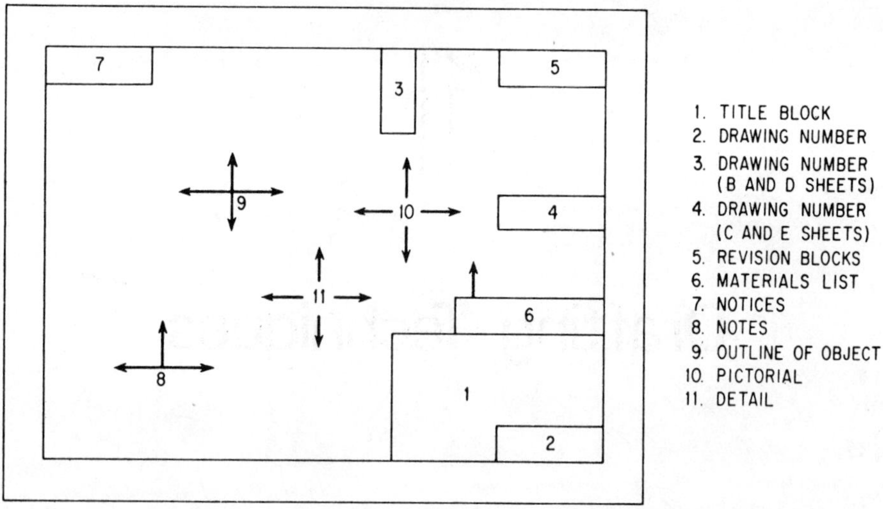

Figure 1 Drawing sheet layout

The general layout of the drawing sheet is shown in Figure 1. For different-size sheets the drawing number will appear at various places on the sheet layout so that when the sheets are folded, the drawing number will appear in the proper place for filing. The location of other features is described briefly in Figure 1 (also see Figures 49, 116, 122, and 159).

Drawing sheets come in sizes, as shown in Table 1-1.

Table 1-1

Sheet	*Society of Automotive Engineers (SAE)*	*Military Standard*
A	9 X 12″	8½ X 11″
B	12 X 18″	11 X 17″
C	18 X 24″	17 X 22″
D	24 X 36″	22 X 34″
E	36 X 48″	34 X 44″

Various types of lines are used in setting up the object and the information on a drawing sheet. The following is a description of the most commonly used lines and their thickness:

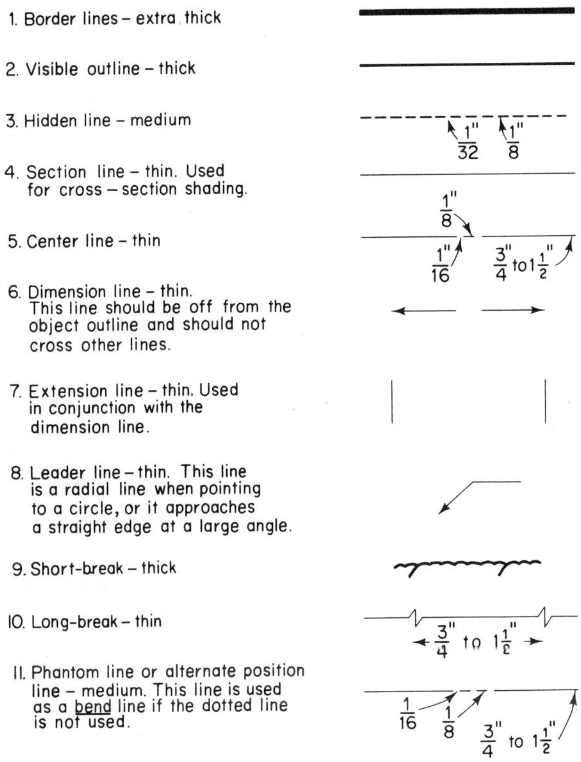

Title block or heading block detail is shown in Figure 2. Either this style of title block or a title block that goes across the entire bottom edge of the drawing may be used. The information provided in the title block should include such items as the following:

The name of the company.

The name of the object.

The signature of the person approving the drawing.

The signature of the person checking the drawing.

The signature of the person doing the drawing.

The scale of the drawing.

The required tolerance of the object.

The date (or various dates).

The drawing number in the lower right-hand corner.

Other miscellaneous information.

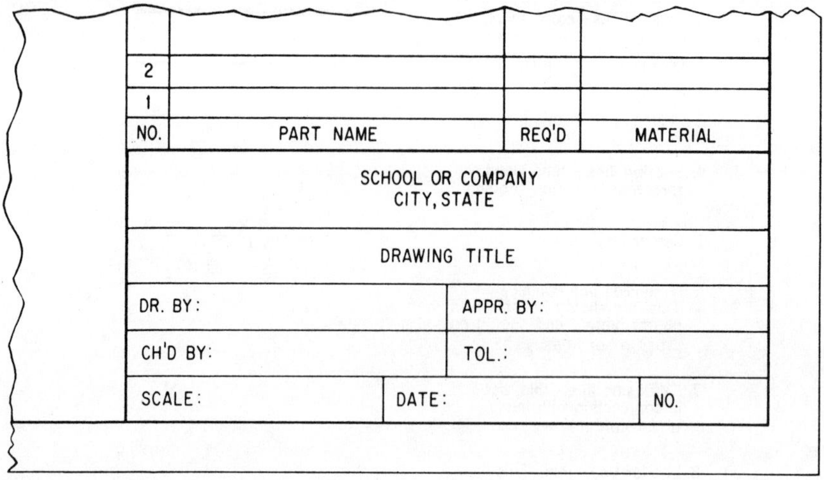

Figure 2 Title block information

The style of title block shown in Figure 2 is used because of its simplicity. Generally the size, shape, and organization of the block may vary and is set up on the basis of common sense. The words printed in sections of the block should not touch the lines and should be centered.

As indicated in Figure 1, the *materials list* is located above the *title block.* Item 1 is placed immediately on top of the title block and the following consecutive numbered items appear in a list built up toward the top of the drawing sheet. The materials list includes the item number, part name, quantity, and description.

It is often not possible to draw the object at a *1 : 1 scale.* When such is the case, the object is either *reduced* or *expanded* in size so that it will fit the paper. This reduction or expansion is indicated in a space in the title block under the heading "Scale:" and with a notation such as 1/2″ = 1″, 2″ = 1″, or 1/8″ = 1″. These notations are interpreted to mean that 1/2″ on the *drawing sheet* is equal to 1″ on the *actual object.* The same meaning holds true for the other examples. The first figure represents distance on the drawing, and the second figure represents distance on the object. Figure 3 shows

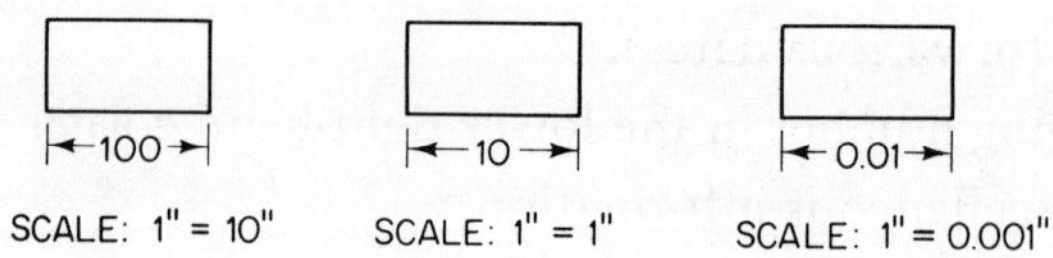

Figure 3 Examples of different drawing scales

examples of using a reduced scale, a 1 : 1 scale, or an expanded scale. Assume that these views are formed on an ordinary 8½ X 11″ size A sheet.

The *tolerance* information on a drawing sheet has to do with how accurately the piece must be made. If it does not have to be made accurately, it has a wide range of tolerance. If it must be made very accurately, it has a narrow tolerance and must be made very carefully. The information about this must appear in the title block under the heading "Tol:" and is written as shown in Figure 4. The person designing the object must decide how accurately it must be made.

Figure 4 Examples of showing the tolerance

Two common systems of dimensioning are used by draftsmen and are pictured in Figure 5. In this figure the dimensioning system shown at "a" has the advantage of being simple, clear, and easy to read. It also is used where a *cumulative error* cannot be tolerated. As can be seen in the illustration, all dimensions are made from a *common reference* edge. The system shown at "b" is the old standby

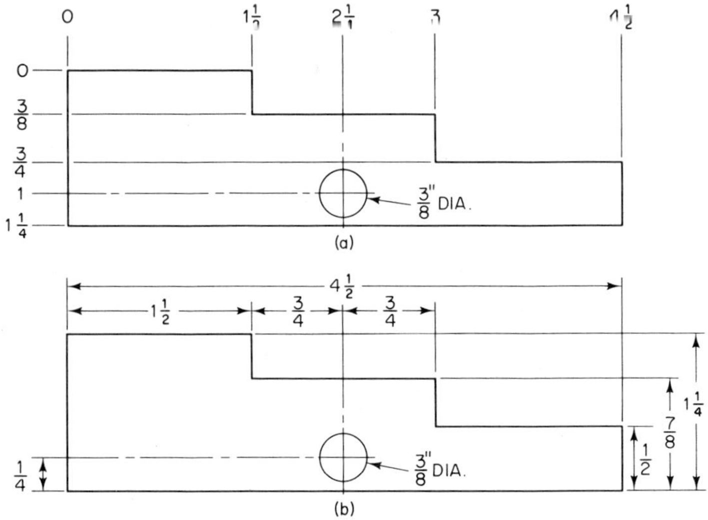

Figure 5 Dimensioning systems

but is still used quite often. In some cases it is too involved and awkward, but it can be used where a common reference is lacking or for the simple indication of a distance between points. In the way the dimensions are set up in "b," a person making the piece would probably work from the left edge, giving somewhat the same effect as using the system shown at "a." The choice of which system to use depends upon the configuration or requirements of the object and perhaps in some cases on individual preference.

The main object (especially a three-dimensional object) is often set up on the drawing sheet to show a top, front, and side view. Flat sheet-metal layouts can be shown in a single view that includes all necessary bending instructions and other construction information (see Figures 49, 116, 122, and 159).

Pictorial drawings are used to clarify the detail of the construction of an object or a portion of the whole piece. These pictorials can be simple line drawings in three-dimensional form, as shown in Figure 6. This type of sketch can be used for making a small representation of the object placed off to the side of a drawing sheet or can be used as the basic arrangement of the main object. The *isometric* view is drawn to a true length. On the *oblique* view the length or depth can vary. On either of these views, guidelines should be used for setting up the proper shape of circles.

On pictorial drawings there are usually *three sets* of lines which make up the view of the object. In each set of lines, the lines are *parallel* to each other. There is a set of vertical lines, a set of horizontal lines for oblique views, and one or two sets of lines that are at some angle. In each set the lines are parallel. It is helpful to

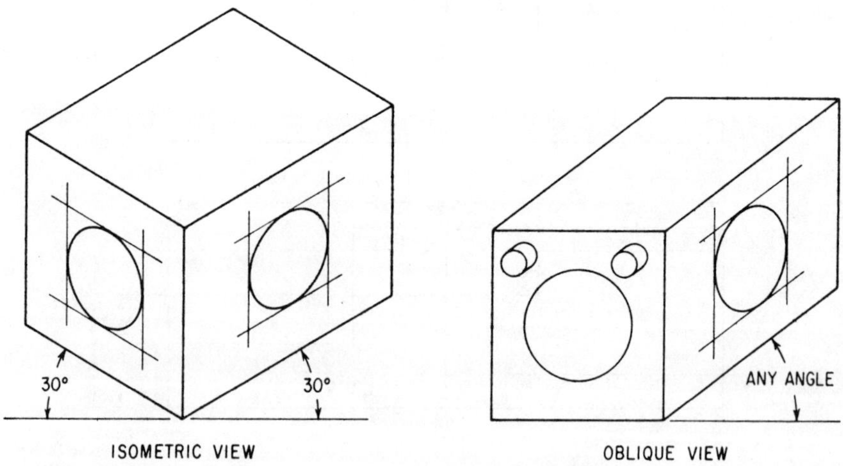

Figure 6 Pictorial views

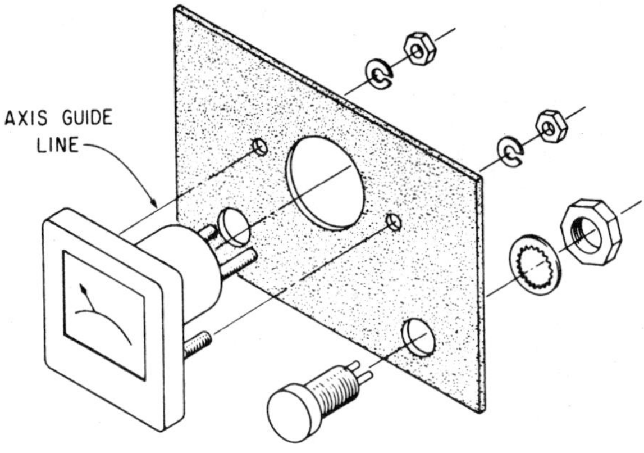

Figure 7 Exploded view

realize this in order to prevent distortion of the view when drawing pictorials.

Another type of pictorial is the *exploded view,* showing the assembly detail of certain parts. A sample of this type of drawing is given in Figure 7. The secret of setting up an exploded view is to use an *axis guideline* for positioning the location of each part. This line is shown in Figure 7.

Wiring pictorials are often used in electronics information manuals. These are simply three-dimensional drawings showing the actual placement of parts and wires. A simple wiring pictorial is shown in Figure 8.

A drawing sheet will usually contain much information. This information must be clear, complete, and spelled out in every detail. At times this detail must be given even at the risk of possible repetition.

Depending upon the purpose of the drawing or the kind of drawing, *nothing* is to be assumed or left up to the other person's knowledge. All the *plans* and *intentions* of the *designer* must be indicated on the drawing sheet, and it cannot be assumed that details will automatically be known by the person making an interpretation of the requirements. All the requirements must be given in a clear and understandable way.

The following items would normally be found on the average drawing sheet:

Outline Information of the Object

1. Overall shape of the piece drawn to scale.
2. Complete dimensions of every part and edge, including special tolerances and bend lines.

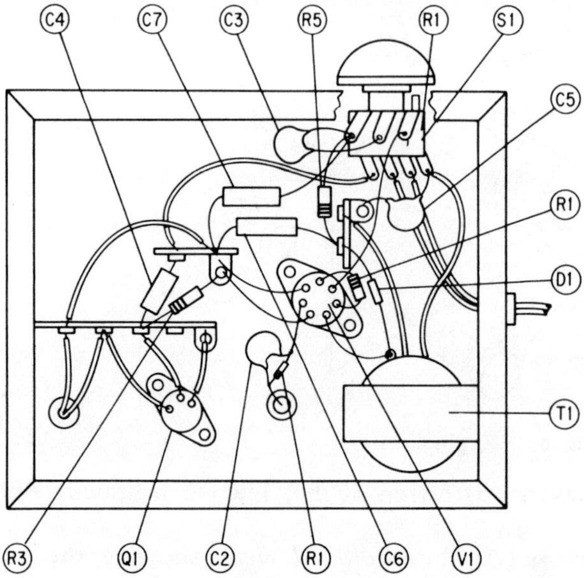

PARTS LIST

C2	0.001 μf
C3	0.02 μf
C4	10 μf, 15 VOLTS, ELECTROLYTIC
C5	0.01 μf, 200 VOLTS, PAPER
C6	10 μf, 15 VOLTS, ELECTROLYTIC
C7	10 μf, 15 VOLTS, ELECTROLYTIC
D1	IN34A DIODE
L1	r-f COIL
Q1	2N321 TRANSISTOR
R1	10 MΩ, 0.5 WATT
R2	1000 Ω POTENTIAL
R3	220 kΩ, 0.5 WATT
R4	1800 Ω, 0.5 WATT
S1	DPST
T1	OUTPUT TRANSFORMER (STANCOR A-3856)
V1	12AE6 TUBE

Figure 8 Wiring pictorial of a simple amplifier

3. Size and location of holes, cutouts, relief cuts, relief holes, ventilation louvers, and depressions.
4. Size of angles or radius lines.
5. Detail information given in an inset drawing, if needed.
6. Three-dimensional pictorials, if needed.
7. Required sectional views.
8. Description of mounted parts.
9. Threaded holes indicated.
10. Labeling information.
11. Hidden detail.
12. Any other special information.

Construction Information

1. Bending, drilling, filing, punching, or cutting information.
2. Spot-welding instructions.
3. Marking or labeling instructions.
4. Instructions for the mounting of parts
5. Order of operations information.

6. Miscellaneous notes.
7. Detail on special operations.
8. Finish information.
9. Any other instructions.
10. Materials list.

Heading Detail

1. Name of company.
2. Name of item.
3. Various signatures and dates.
4. Drawing scale, tolerance, and special requirements.
5. Drawing number.
6. Revisions or notices, if they apply.

In summary, common sense and logic are the basis of good readable drawings. Keep the drawings from being cluttered and disorganized. Whenever possible, use simplifications such as the abbreviation "Typ." in cases of repetition. Keep the object as clear and easy to see as possible, with dimensions and notes on the outside of its outline. Include *all* necessary information on the drawing sheet. *Nothing* should be taken for granted as being understood by the person reading the drawing. Use very fine guidelines to aid in printing the lettering. The dimension numbers should be right side up when viewed with the drawing sheet held in the normal position.

It is suggested that drawings should be judged on the basis of neatness, line rendering, dimensioning, title block information, composition, and completeness.

It is presumed that most students have had at least a small amount of drawing experience. Extra study, research, or practice will be needed by those who have had limited instruction.

Purpose of Assignment

1. To learn basic information related to a line drawing.
2. To gain experience in planning the arrangement of a drawing sheet.
3. To gain experience in making a comprehensive set of plans that is complete in every detail.
4. To learn to work carefully and neatly in making a finished drawing.

Tools and Equipment

No. 2 lead pencil.

Ruler with straight edge.

Soft eraser.

Materials

Drawing paper, size A (8½ × 11″).

Assignment

1. Make a flat layout drawing of the bracket shown in Figure 9, or the component shelf of Figure 10. Make the drawing complete in every respect.
   a. Include at least three holes or cutouts if you draw the component shelf and draw only the component shelf piece and not the front panel.
   b. Determine and give *all* dimensions and other necessary construction information.
   c. Choose *one* of the two systems of giving dimensions.
2. Make an isometric or oblique pictorial view of a simple three-dimensional object assigned by the instructor.
   a. Dimensions possibly would not be a part of this type of

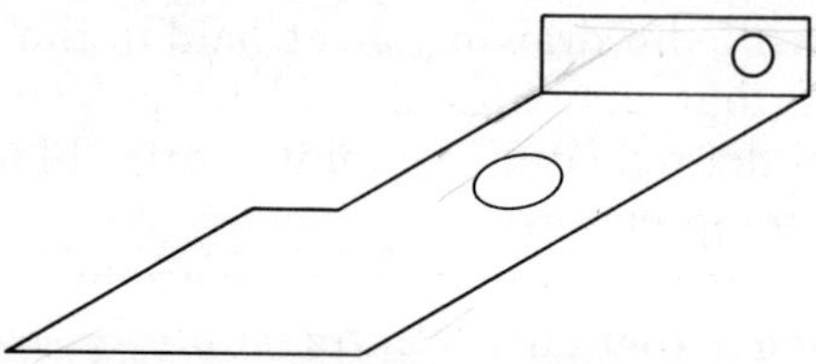

Figure 9 Pictorial drawing of bracket

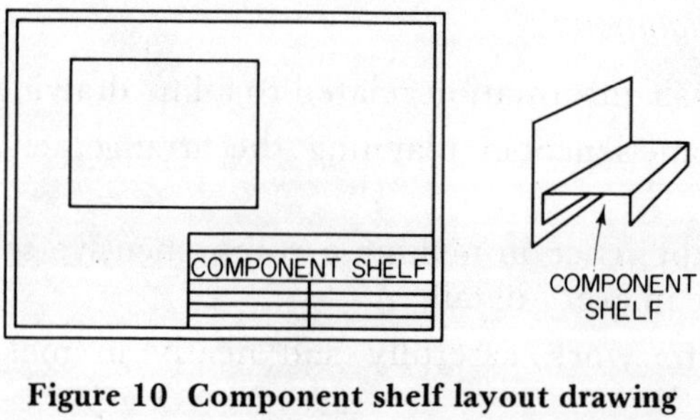

Figure 10 Component shelf layout drawing

drawing, but it is important to make the drawing *proportionately* correct.

b. The drawing would include a border line, a title block, and any other information that would relate to the purpose of the drawing.

Directions

1. Make up all the drawings on 8½ X 11″ sheets of paper. Use a ruler and a No. 2 pencil for making all the lines. Use very lightweight guidelines for setting up the basic arrangement of the drawing sheet.
2. Refer to earlier paragraphs in this chapter for information on line rendering and the drawing sheet layout.
3. Use your imagination or good judgment from past experience for determining the number, size and location of holes, or other construction detail.
4. Determine *all* the information necessary to complete the drawings. Do not leave anything out. Include *all* necessary instructions and any information that would normally be found on a drawing sheet.
5. On the pictorial drawing be sure to keep the correct proportions of height, width, and depth of the object and also any detail that is involved.
6. As the drawings are assigned, complete them and turn them in to your instructor.

2

Assembly Techniques

INTRODUCTION TO SOLDERING (1)

The use of solder in the electronics industry is a tremendously important solution to a great number of potential problems. Solder provides an inexpensive and fast method of joining two or more metals together without the need for costly tooling or elaborate equipment. A completely metallic and excellent conductive medium is obtained by the use of solder, resulting in a highly stable electrical continuity. The soldered connection will also withstand physical stresses, strain, vibration, expansion and contraction due to temperature changes, and corrosion, and can be used for liquid or airtight seals. Connections in electrical equipment can be made without the use of solder, but in some ways the easiest high-quality connection is the soldered one. Knowledge of a few basic requirements and good technique is all that is necessary to do good soldering.

Solder makes an attachment to copper by actually dissolving a small amount of the copper and is thereby fused into the surface of the copper. This is a metal-solvent action and is chemical in nature.

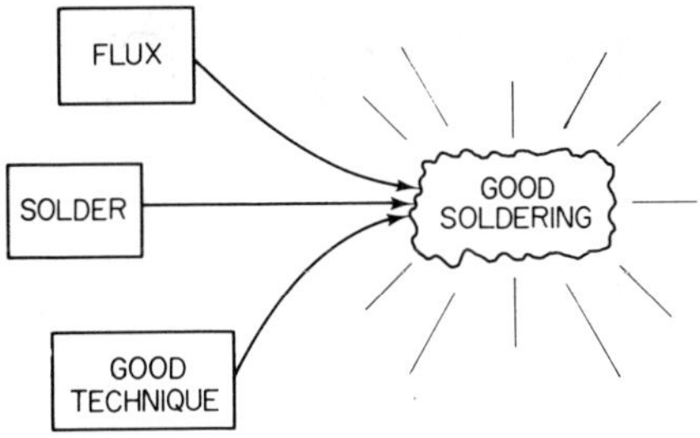

A new alloy is formed at the junction and there is an unbroken metal continuity, which is ideal in terms of an electric connection. When two metals are soldered together, they act like one metal (see Figure 11).

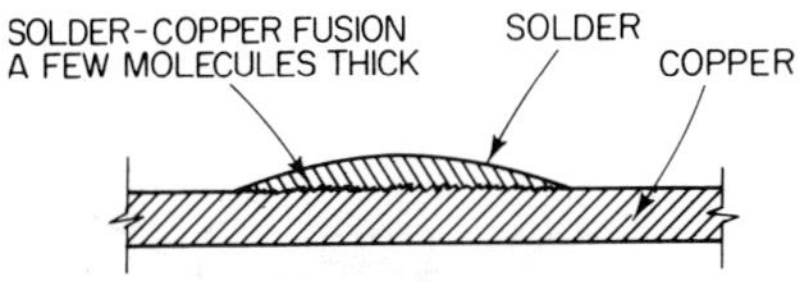

Figure 11 Solder-copper fusion

Most of the soldered connections in electrical equipment involve metal conductors of one kind or another, which in turn are soldered to a terminal or connection point. The conductor most commonly found will be stranded wire, single-strand solid wire, copper buss bars, or copper foil on printed-circuit boards. Often these wires or conductors will be pretinned with a coating of tin or solder, silver-plated, or plated with some other special metal. The terminals or connection points will vary in basic shape and structure. Generally, the types most commonly found will be the turret, flat eyelet, split, bifurcated, hook terminals, and the pad or land found on the printed-circuit board. See Figure 12 for a view of some of the commonly used terminals.

To partially eliminate heat problems during soldering and also to simplify parts-replacement problems, the wires and components are to be placed on a terminal in the proper order. The sequence, as shown in Figure 13, is as follows:

1. Jumpers between terminals.
2. Jumpers to nonadjacent terminals.
3. Loose wiring.
4. Harness wiring.
5. Small components.
6. Heat-sensitive semiconductors or components.

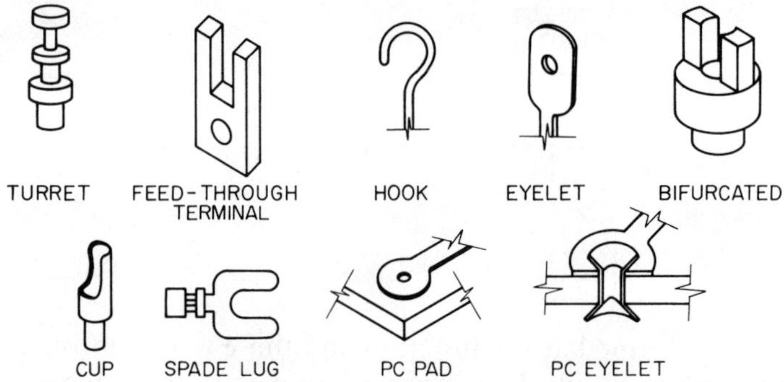

Figure 12 Commonly used terminals

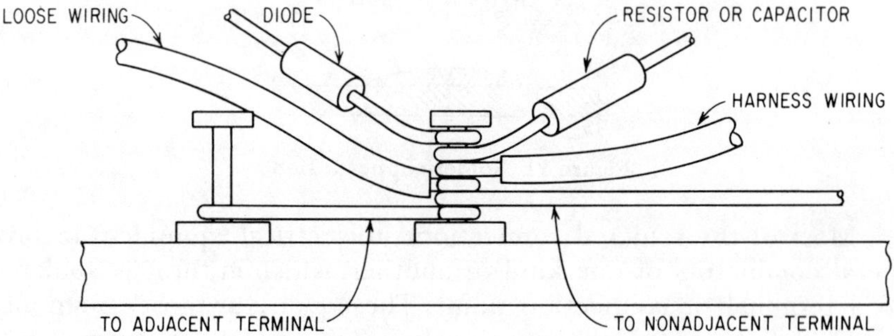

Figure 13 Terminal-wiring sequence

The following is a basic list of tools required for soldering:

1. Ritchie, *Electronics Construction and Assembly.*
2. Soldering device, 40- to 50-W.
   Unger—handle #777 or 776.
   Heating element #1235 and tip #PL333 (or equivalent).
3. Long-nose pliers—4 to 6″.
4. Diagonal cutters—4 to 6″, full flush cutting type, *or* regular diagonals and straight-edge nail clippers.
5. Soldering vise.
6. Screwdriver—3″, flat blade.
7. Heat sink or antiwicking tool.
8. Scribe (can be homemade from old compass point).
9. 6″ steel rule.
10. Small cleaning board (homemade from Masonite or Formica, 4 × 5″).

11. Braid component lead cleaner (see *Coaxial-Cable Breakout* assignment).
12. Ink eraser.
13. Cleaning brush.
14. Safety glasses for eye protection.
15. Wiping rag—hard-weave and lint-free.

Some of the tools listed above are shown in Figure 14.

The list above is not a complete tabulation of all tools used in soldering but rather a list of the basic tools that would be required by a person starting out to gain experience in the field of electronics assembly. Most of these tools should be considered by students to be essential for developing proper soldering and assembly techniques.

Hand-held soldering irons come in various sizes and styles. Many have replaceable tips, allowing them to be used in specialized applications. For soldering the average connection or terminal, a 40- to 50-W iron is adequate and will give sufficient heat in a short period of time. For heavier work, of course, a 100-W iron may need to be used. Following proper preparation, the soldering operation is performed very quickly in order to prevent damage to components, burned insulation, and solder wicking. This means that selection of the proper soldering tool is important, that proper care and preparation of the tool is necessary, and that a workable technique must be developed.

The quick-heating solder gun is not the best tool to use for

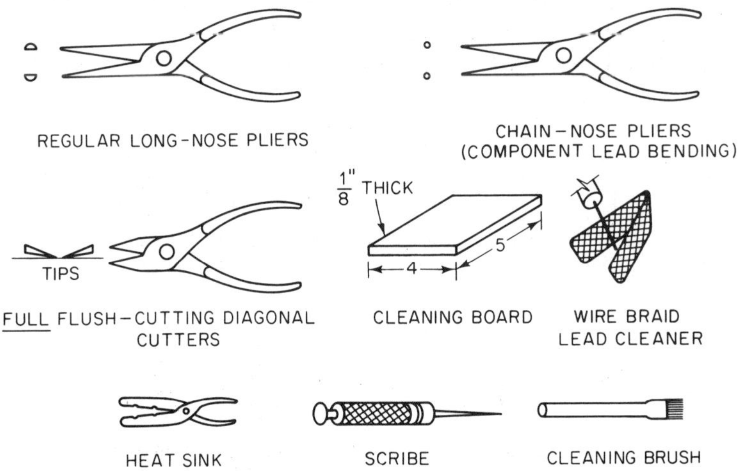

Figure 14 Soldering tools and equipment

beginning solder practice. This tool works well in a repair situation where one or two connections need to be made quickly, but it is difficult to make the proper application of heat when doing carefully inspected soldering.

Many specialized types of soldering tips are available, the most common being the pyramid, the chisel, and the pencil-tip. Generally, the shape of the tip is not critical, and with a slight preference toward the pyramid point, either the pyramid or chisel can be used equally well. The pencil-tip should be used for fine or delicate work. The various tips are constructed of solid copper or of other materials that are plated for protection against corrosion. Depending upon the type of construction, each type of tip requires a special procedure for keeping it clean and usable. The different types of tips have desirable or undesirable features, and the selection, of course, would have to be made on the basis of individual requirements. Generally, the solid copper tip is easy to use except where the iron is left on for long periods of time. The plated tip is better able to resist the deleterious effects of heat and should be used when the iron is kept heated over extended periods. See Figure 15 for a view of various shapes used on soldering iron tips.

The procedure for keeping the tips clean during the soldering operation will depend upon the type of tip used. The solid copper tip should be coated (tinned) with solder when it is first turned on, and periodically recoated. It can be wiped from time to time with a hard-weave cloth to remove the accumulation of dirt or oxides. When solder is added to the tip and also during the soldering operation, the flux in a rosin-core solder will aid in keeping the faces on the end of the tip tinned. When the tip gets rough or pitted, it may need to be filed with a flat file and recoated with solder.

The plated tips are cared for during their use by continually

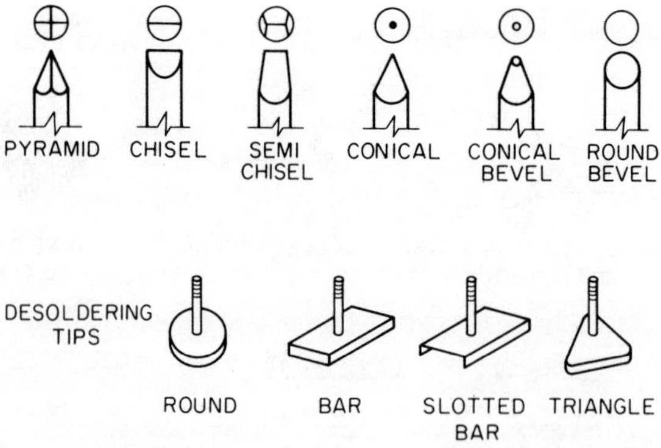

Figure 15 Soldering iron tip shapes

wiping them with a rag or wet sponge and at the same time tinning the faces of the tip with a coating of solder. A plated tip *must never be filed.* This would obviously destroy the plating and eliminate the advantages that are incorporated in this type of tip.

The solder itself is an alloy of tin and lead. These will be found in different percentages, depending upon the desired characteristic of the solder. A ratio of these percentages is used for identification of the alloy. The solder used commonly in electrical connections is the 60/40 and in some cases 50/50. The first number of the ratio represents the percentage of tin and the second number represents the percentage of lead. Tin melts at 450°F, lead melts at 621°F, and interestingly enough, a 60/40 solder melts at 375°F, which is a lower temperature than the melting point of either of the two metals making up the alloy.

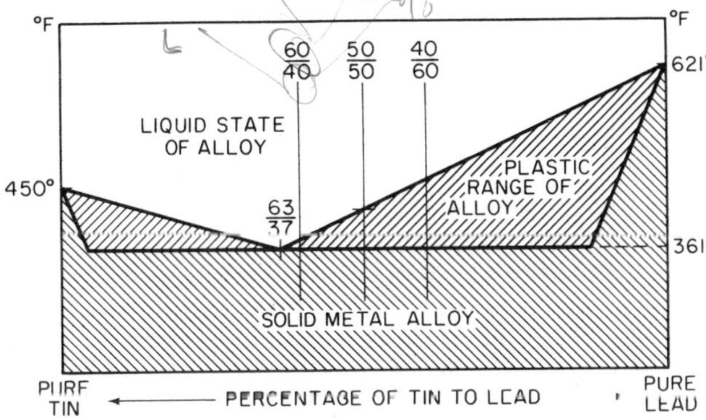

Figure 16 Melting characteristics of solder

Depending upon the percentages of tin and lead, the characteristics of the resultant alloy will vary considerably. These different characteristics can be shown graphically, and it can be seen that the melting point and plastic range (semimolten state) of the solder will be most affected as the tin/lead combination is changed. With a combination of 63/37, the alloy has its lowest melting point and narrowest plastic range, and both of these factors will increase as the percentages are changed in either direction. Refer to Figure 16.

It is recommended that the 60/40 solder combination be used. This solder melts at a low temperature (375°F) and has a narrow plastic range, a high quality of fusion with the base metal, and greater stress resistance than most other alloys.

Another important factor directly related to good soldering is the use of the proper flux. No matter how clean the surface of the metal to be soldered might be, it is necessary to have further cleaning action so that metal-solvent action between the solder and copper can take place. This is accomplished by the use of a rosin or resin flux. This type of flux is active in dissolving oxides and contaminants, when heated, but is perfectly inactive when left as a residue. Flux melts onto the surface before the solder melts, does its job of cleaning, and then is pushed up out of the way to be replaced by the solder. The flux has a wetting action, prevents or cleans oxides, and breaks down the surface tension of the solder. This allows the solder to spread out smoothly onto the metal without balling up or only partially adhering to the surface. Flux is usually applied automatically by virtue of the fact that it is contained as a core (or several cores) inside the solder-wire strand. The size of the core comes as a certain percentage of the strand size, which determines the respective amount of flux applied. Core size is designated as regular, medium, small, tiny, or multicore. *No acid-core solder* should ever be used around or on electric wiring. Always be positive that the flux is a rosin flux (see Figure 17).

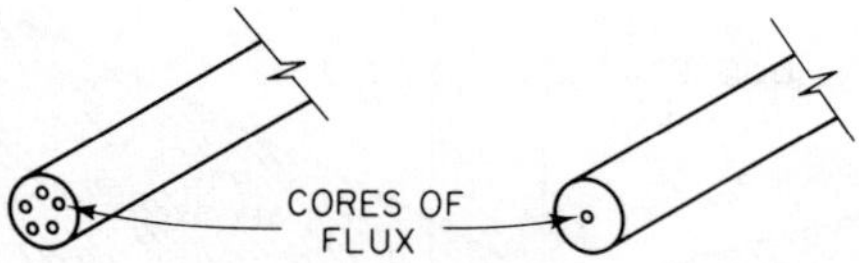

Figure 17 Cross section of solder wire

Solder wire comes in various wire-gauge sizes. Normally, a 16-gauge to 22-gauge wire is used, depending upon the type of work being done. If more solder is required in a short period of time, the 16-gauge strand should be used. If only a very small amount of solder is needed, the 22-gauge strand should be used. When ordering solder, specify the strand size, alloy, core size, and type of flux.

After the terminal is prepared, there are several ways of applying the solder to make the connection complete. Among others, these methods include the use of a hand-held soldering iron, resistance soldering, dip soldering, machine soldering, percolation, and induction soldering. The various methods are adapted to meet individual requirements and, in many cases, a production-line situation. The hand-held soldering iron is still an important tool applicable to small production or piece work. Soldering techniques relating to the use of this tool will be emphasized in this assignment.

Prior to making the connection of a wire to a terminal, unless a bare-wire jumper is used, the insulation on the end of the wire must be stripped back a certain distance. On the production line this distance would be determined by the requirements to be fulfilled, and the stripping of the insulation would be done on a mass-production basis. Where rigid requirements prevail, a thermal wire stripper is used;

the use of the mechanical stripper, which can nick the conductor, is in many cases not allowed. Used with a special type of insulation, the thermal stripper has a loop of wire that heats up, melting a ring around the end of the wire. This allows the insulation to be slipped off easily. Various types of wires would be handled according to their particular characteristics.

For hand wiring it is recommended that the wire be stripped back about 1/2 to 5/8". This allows sufficient conductor to be used for wrapping around the terminal, and any excess can easily be clipped off at a later stage of the operation. Before actually wrapping the conductor around the terminal, it is difficult to judge the exact amount of conductor length that will be needed. But just prior to the completion of the operation, it is a simple matter to clip the end of the conductor to the exact length. This will be described in greater detail in the soldering operation sequence given later in this introductory section.

A typical terminal and wire wrap are shown in Figure 18. Good wiring technique requires that there be a good mechanical connection prior to the application of solder. A right-hand wrap should be used whenever possible and the leads should not overlap when being wrapped around the terminal. On multiwire terminals the leads should be wrapped in the same direction around the terminal. It is further specified that the wrap around the terminal be not less than three-fourths or more than one turn to allow for several connections to the same terminal. Some military specifications require a 1/16" pigtail on terminal wraps and others do not permit it. Unless the pigtail is called for, the specification given above should be used as a guide.

Components should be mounted so that identification or polarity is readable. To prevent the shorting of long exposed leads, sleeving

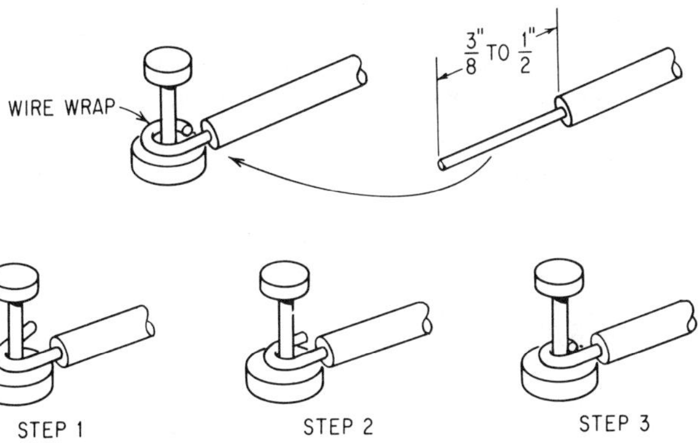

Figure 18 Connecting wire to a terminal

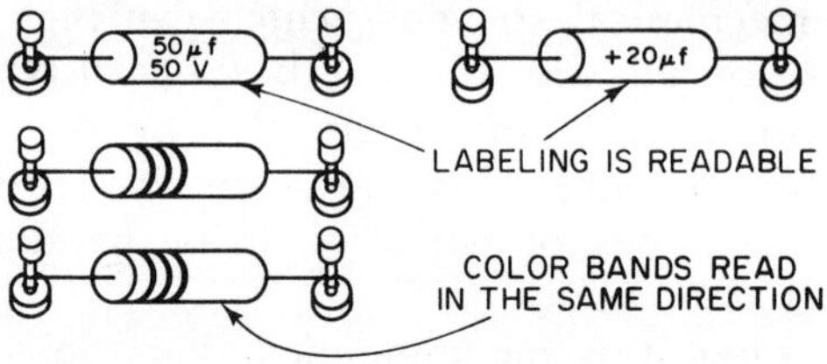

Figure 19 Component orientation

must often be used. For strain relief there should be a certain slack in jumper wires or component leads. Figure 19 shows components oriented properly between terminals.

During the actual soldering operation, care must be exercised to prevent certain undesirable results. If the component leads to be soldered have not been heated enough, a *cold solder joint* is the result. In this type of joint the solder does not melt sufficiently and flow out smoothly. The *fractured joint* is one in which some sort of a movement has caused the parts of the joint to move while the solder is in its plastic state. At this time the solder is neither solid nor liquid, and any movement will cause it to "crystallize," break apart, and form a very rough connection. A *rosin joint* is one in which a part of the joint has been heated enough to become coated with flux but not sufficiently to melt the solder. This part of the joint is left coated with rosin, which acts as an insulator between that part and the rest of the connection.

Wicking or solder creep is an effect that must be guarded against when soldering stranded wire. Solder follows heat and if a terminal or wire is heated too much, the solder will follow the heat back under the insulation on the wire and cause the wire to be changed from a stranded to a solid form of wire. Vibration will then more likely cause the wire to break at the connection point. Needless to say, the application of excessive heat will also cause the insulation to burn or melt. Refer to Figure 20 for information on tinning a stranded wire.

To adequately control the soldering problems mentioned above,

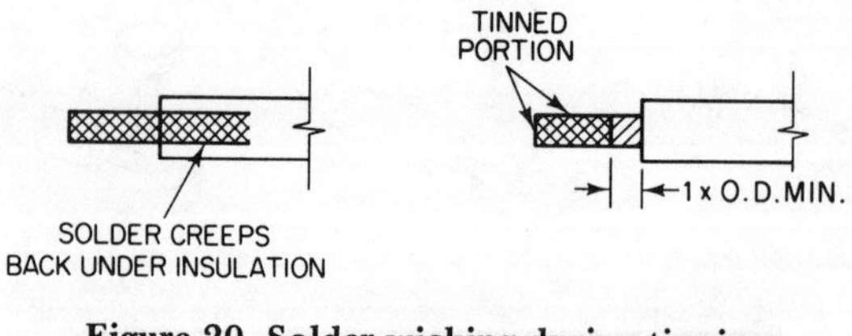

Figure 20 Solder wicking during tinning

it is extremely important to work with a clean soldering iron and clean workpieces. It is necessary to have sufficient heat on the parts to be soldered, and it is necessary that this heat not be allowed to damage the parts that are involved. These problems can be taken care of by supplying sufficient heat to the work quickly, and completing the whole operation in a *short* period of time. The heat will transfer quickly from the soldering-iron tip to the work if the tip is clean and tinned so that a "bridge" of solder can be formed at the point of contact. Sometimes a small amount of solder may need to be added to form this bridge (refer to Figure 21).

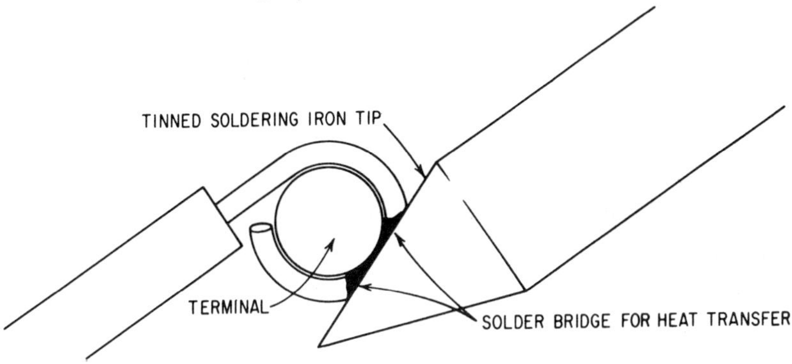

Figure 21 Soldering a terminal

The general sequence of steps involved in the soldering of a terminal is as follows:

1. *Prepare the iron.* Heat the iron; clean and tin the tip. Leave the tip coated with a good coating of solder.
2. *Prepare the work that is to be soldered.* Remove old wires from the terminals. Remove all excess used solder and old flux. Wipe clean and dry as required. Wear safety glasses.
3. *Prepare the new leads.* Cut the new leads to length. Strip the insulation. Tin the conductors as required.
4. *Connect the leads to the terminals.* Adjust the insulation distance. Form and wrap the wires around the terminals. Make a tight mechanical connection slightly less than one full turn around the terminal. Attach the heat sink as required.
5. *Solder the connection.* Clean the tip of the iron with a sponge. Solder the connection using flux. Apply the proper amount of solder to the connection.
6. *Clean the work.* Clean all flux residue from terminals and board.
7. *Check the work.* Correct connections will be uniform in appearance.

The following items would be considered in an inspection of the soldered connection:

Nicked or partially cut leads.

Too much solder or not enough solder.

Burned insulation.

Insulation back too far or pulled in too close.

Leads not properly crimped close to the joint.

Component leads too long.

Fine wire strands not soldered.

Leads not cleaned or leads improperly soldered.

Leads too short and under strain.

Solder splashes.

Solder wicking (solder creep on stranded wire).

Fractured joint.

Rosin joint.

Cold solder joint.

In addition to the items above, soldered connections can be judged on the basis of *uniformity*. If the basic standards are followed, high-quality connections will be found to be quite uniform in their appearance.

Purpose of Assignment

1. To learn the basic standards of good soldered connections and terminal wiring.
2. To learn the characteristics of solder and solder flux.
3. To learn about soldering devices and safety procedures.

Tools and Equipment

Refer to the list of tools given earlier in the chapter.

Materials

18-gauge solid hook-up wire.

18- or 20-gauge stranded hook-up wire.

22-gauge rosin-core solder (60/40).

Liquid soldering flux.

Board with mounted terminals and miscellaneous components.

Octal tube socket mounted on a board or equivalent type of terminal.

Flux remover, isopropyl alcohol, or Freon vapor degreaser.

Assignment

1. Using the solid hook-up wire, make several single connections to the terminals on a terminal board.
2. Using the stranded hook-up wire, make a connection to each of the terminals on an octal socket.
3. Solder miscellaneous components to terminals on a terminal board.

Directions

To prevent damage to the associated parts, the actual application of heat and solder to a connection must be accomplished very quickly and without delay. A heat sink should be used with stranded wire and on delicate parts, and only the proper amount of solder should be added. Generally, the contour of the wires can be seen in a properly soldered connection. A rosin-core 60/40 solder is recommended.

Solid-Wire Connections

1. Turn on the soldering iron. Coat the faces of the tip as soon as they are hot enough to melt the solder. Use a soldering iron with a plated tip. Wipe the faces of the tip on a moistened sponge. *Do not file* a plated tip. When the tip is cleaned and tinned with an adequate amount of solder, place the iron on a soldering stand.
2. Prepare the terminal board for making the wire connections. Clean old wires off the terminals and remove excess solder by heating the terminals and wiping or brushing. Be sure that all dirt and excess solder is removed. In some cases new terminals should be tinned and in all cases be sure that the terminals are clean. Add flux to aid in cleaning the terminals and *avoid solder splashing in your eyes.*
3. Prepare the solid wire for wrapping around the terminals. Remove the insulation by nicking it in the proper place and prying it off with the diagonal cutters. Use wire strippers if desired. In either case, be careful not to nick the wire conductor.
4. Wrap the conductor around the terminal. Hold the conductor in its proper position and bend the end of the conductor partway around the terminal using the long-nose pliers. Be sure that

the end of the insulation is the proper distance away from the terminal. This should be a distance equal to the outside diameter of the insulation itself. Refer to Figure 26.

5. Clip the end of the conductor so that it will wrap slightly less than one full turn around the terminal and crimp the remaining end in close and tight to the terminal. Add flux to the connection. See Figure 21.
6. Wipe the tip of the soldering iron with a sponge or wiping rag. Hold the tip of the iron against the terminal and be sure that a bridge of solder forms a path for the conduction of heat to the terminal. Solder may need to be added between the terminal and the tip to form this bridge (see Figure 21). Always *be careful not to burn yourself* with the soldering iron.
7. Almost instantly following step 6, touch the solder to the wire or terminal and let the proper amount of solder melt into the connection.
8. Remove the iron and allow the connection to cool without moving or disturbing the solder until it solidifies.
9. Clean off the flux residue with a flux solvent such as flux remover, isopropyl alcohol, or Freon vapor degreaser. *Do not splash solutions* into your eyes.
10. Complete the connection of wires to several terminals and inspect each connection for correctness. Look for uniformity and consistency between connections.

Stranded-Wire Connections

1. Follow the directions for connecting solid hook-up wire to terminals.
2. Twist the strands of the conductor together tightly after the insulation has been removed. Apply liquid flux to the stripped end of the wire. Use a heat sink and tin the wire (refer to Figures 31 and 32).
3. Wrap the tinned conductor around the terminal to which the connection is to be made. Be sure that the wire only wraps slightly less than one full turn around the terminal and that the insulation is the proper distance from the terminal.
4. Attach a heat sink on the metal part of the wire between the end of the insulation and the terminal. Add flux to the connection. Refer to Figure 33.
5. Clean the tip of the iron and solder the connection. Perform the actual soldering operation very quickly to prevent solder creep.

Have everything ready and be sure there is no delay at the time the heat is applied. Be sure there is a good solder bridge for a rapid transfer of heat to the socket terminals and the wire connections.

6. Run the leads to the socket terminals along the mounting board and up to the terminals. Do not loop the wire connections up in the air and then to the octal socket terminals.
7. Complete eight *perfect* connections to the octal socket.
8. Evaluate the connections on the basis of the items listed earlier in the chapter. If the connections are not neat and uniform, determine in what way they differ. Remake the connections if necessary.
9. Clean off the flux residue using alcohol, flux remover, or Freon vapor degreaser.

WARNING: Remember to observe safety precautions in all phases of the soldering operation. *Above all,* be sure that no molten solder gets into your eyes. If this does occur, it is very likely that *permanent* damage will be done.

SOLID-WIRE TINNING AND SOLDERING (1)

Wire with a varnish insulation is very often used for winding coils and transformers. This insulation is usually a tough coating of polyvinyl resin, polyurethane films, or glass fibers bonded with silicone. As electrical insulation it can withstand a certain amount of physical punishment and temperatures up to approximately 180°C. This sort of insulation also has the characteristic of being very thin compared to other types of insulation and can be used where it is important that the insulation occupy as little space as possible.

Teflon, plastics, rubber, and fibers are also used as wire insulation. The characteristics of these insulations vary depending upon the material being used. In general, most of these insulations are rather thick compared to the varnish insulation described above. Also, they are good only for low temperatures and can be easily damaged by heat.

Most wires used in electrical work are prepared for use by removing at least a small portion of the insulation at the end of the wire where the connection will be made. Generally, the insulation is removed by using mechanical wire strippers, a thermal stripper that melts off a portion of the insulation, or by scraping the end of the wire, as in the case of the varnish insulation mentioned above.

When insulated wire connections are made to a terminal, the end of the insulation should be carefully adjusted so that the clearance from the terminal to the end of the insulation is approximately equal to the outside diameter of the insulation itself. This is a standard rule and is shown in Figure 22.

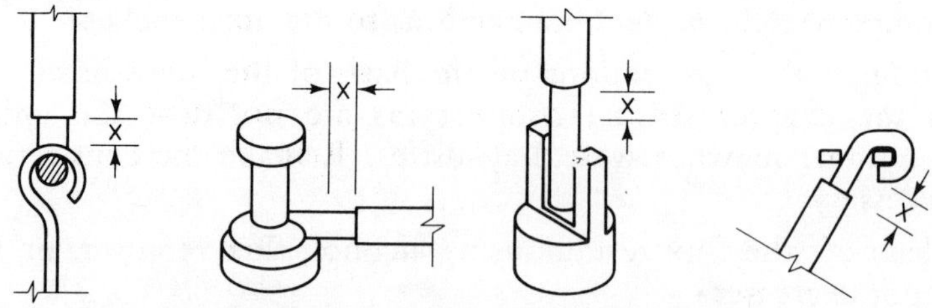

Figure 22 Insulation clearance at a terminal

Purpose of Assignment

1. To gain experience in scraping varnish insulation and preparing a solid wire for soldering.
2. To learn how to tin a heavy-gauge solid wire.
3. To solder a heavy-gauge solid wire to a terminal by controlling the heat.
4. To solder a solid hook-up wire to a terminal.
5. To learn proper safety procedures when soldering.

Tools and Equipment

Various hand tools for soldering.

Soldering device with dampened sponge.

Soldering vise.

Air hose.

Safety glasses for eye protection.

Materials

9″ piece of 12-gauge solid copper wire with varnish insulation.

9″ piece of 20-gauge solid wire with plastic insulation.

Liquid flux and isopropyl alcohol, flux remover, or Freon vapor degreaser.

Terminal board with feed-through terminals.

22-gauge rosin-core solder.

Assignment

1. Tin both ends of two pieces of 12-gauge wire 4″ long (four ends tinned).
2. Solder one end of each 12-gauge wire to a feed-through terminal (two solder connections).
3. Solder both ends of two 4″ 20-gauge solid-wire jumpers to four terminals (four solder connections).

Directions

For information on soldering, refer to the *Introduction to Soldering* section at the beginning of the chapter.

12-Gauge Wire Connections to a Terminal

1. Cut the 12-gauge wire into two pieces exactly 4″ in length.
2. Scrape the varnish insulation from both wires to exactly 3/8″ back from the ends. Use the side edge of a flat-blade screwdriver as shown in Figure 23. Slant the screwdriver a small amount as shown and scrape the wires carefully so that there is an even, smooth edge where the varnish insulation stops. The scraping can be done with the wire held on the left forefinger and the screwdriver moved outward from the body, or with the wire between the right thumb and the screwdriver, which is held with the right-hand fingers.
3. Add flux to the bare copper ends of the wires and tin each end with a thin, smooth coating of solder. The soldering device can be held in the vise as shown in Figure 24.
4. Place one end of each wire into a feed-through terminal with the wire protruding 1/8″ through the terminal.
5. Add flux to the connections and solder the wires to the terminals

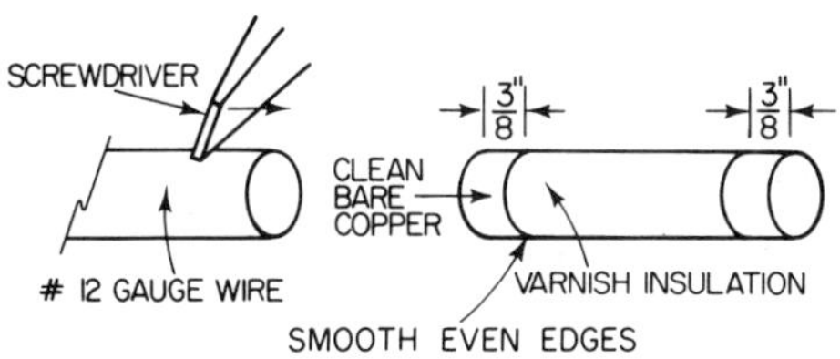

Figure 23 Scraping varnish from wire

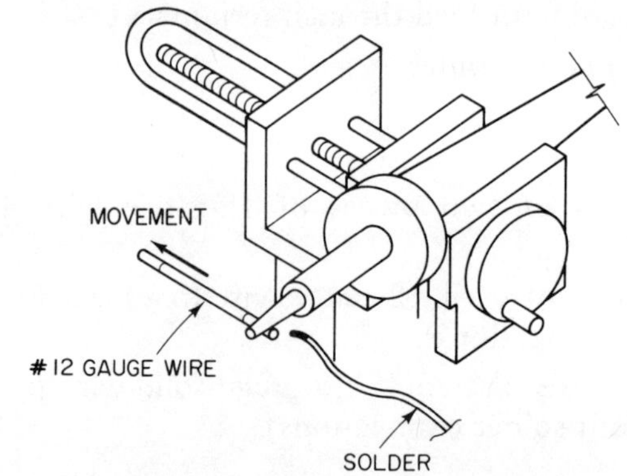

Figure 24 Tinning the heavy-gauge wire

with a smooth fillet of solder ending up between each wire and its terminal as shown in Figure 25. Control the application of heat carefully. Too little heat will result in very rough connection, and too much heat will cause most of the solder to flow away from the joint. Work safely.

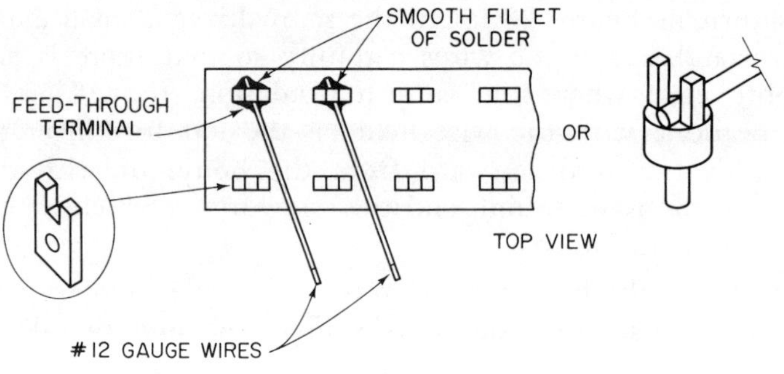

Figure 25 Heavy-gauge wire connections

20-Gauge Wire Connections to a Terminal

1. Cut the 20-gauge wire into two pieces exactly 4″ in length.
2. Using the wire strippers, strip the insulation 5/8″ back from the ends of the wires.
3. Using the diagonal cutters, cut any threads or rough edges of the insulation to make it smooth at the ends.
4. Connect one end of the 20-gauge hook-up wire to a terminal. Note the following steps, as shown in Figure 26:

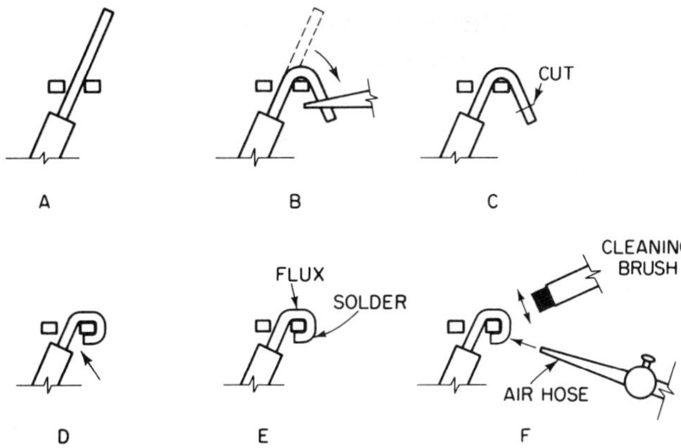

Figure 26 Hook-up wire connections

a. Place conductor against terminal with the insulation at a proper distance from the terminal.
b. Wrap the conductor 3/4 turn around the terminal. Use the long-nose pliers.
c. Cut off the excess end of the conductor, judging the length to make it slightly less than one full turn.
d. Crimp the end of the conductor in tight against terminal. Use long-nose pliers.
e. Apply flux and solder the connection.
f. Clean the terminal with alcohol or flux remover and dry with an air hose, or use Freon vapor degreaser.

5. Complete the connection of each end of each wire to a terminal (four connections with 20-gauge solid wire). Keep safety procedures in mind.

Cleaning the Board After Soldering

1. Clean the flux residue from the terminals by one of the following methods:
   a. Brush with isopropyl alchohol and dry. Remember eye protection.
   b. Dip the piece in a bath of flux remover and dry with an air hose.
   c. Spray with Freon using the vapor degreaser.
2. Put your name on the terminal board and turn it in to your instructor.

STRANDED-WIRE TINNING AND SOLDERING (1)

Some of the most difficult soldering problems arise when making proper high-quality connections using certain types of stranded wire. In these cases problems can develop involving solder wicking, insulation damage, frayed strands, and difficulty in making a proper wrap of the conductor around the terminal.

Solder connections can be made by pretinning the parts that are to be soldered together, or by simply soldering the connection after the parts have been put together. Pretinning has the advantages of quick, easy solder flow and control of the amount of solder. In the case of stranded wire, tinning is done to bond the strands of the conductor together to prevent fraying during handling.

For the average type of connection the conductor is wrapped around the terminal slightly less than one full turn. This is shown in Figure 27. Notice that there is a good, close mechanical connection of the conductor to the terminal and that the conductor end is cut just short of making a full turn around the terminal.

When insulated wire connections are made to a terminal, the end of the insulation should be carefully adjusted so that the clearance from the terminal to the end of the insulation is approximately equal to the outside diameter of the insulation itself. This is a standard rule and is shown in Figure 28.

Although a rosin-core solder is used in hand-soldering processes, it is highly recommended that in most soldering operations extra flux should also be added to the connection prior to soldering. Usually, sufficient flux does not come out of core solder to cover the connection. The external application of a drop of flux greatly enhances solder flow and therefore should be considered necessary when quality soldering is the desired result.

Solder wicking (solder creep) is also a problem when working with stranded wire. Solder follows heat and heat follows conductors.

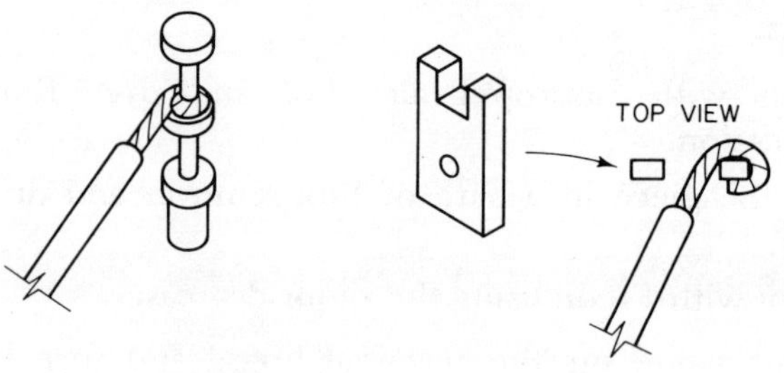

Figure 27 Terminal wire wrap

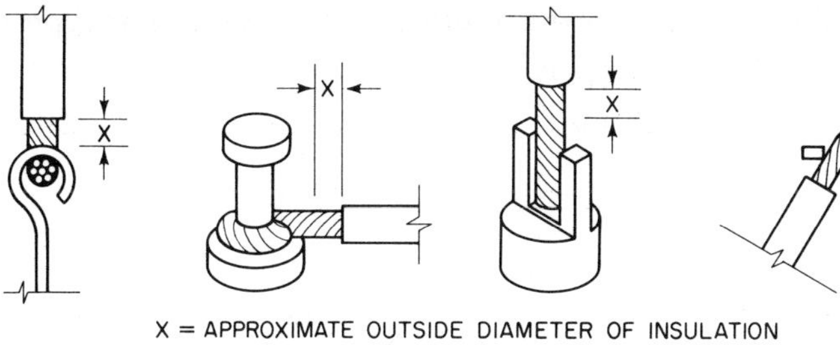

Figure 28 Insulation clearance

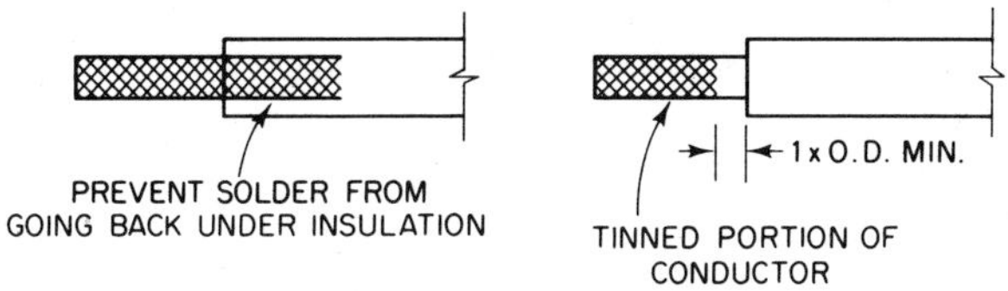

Figure 29 Solder wicking

Too much heat applied to a wire can result in melted insulation and solder creeping up under the insulation as far as 3/4″, as shown in Figure 29. This, in effect, makes the stranded wire into a solid wire which is not flexible and which is apt to break.

Purpose of Assignment

1. To learn how to properly strip the insulation from stranded wire.
2. To learn how to tin stranded wire.
3. To gain experience soldering stranded wire to a terminal.
4. To learn proper safety procedures when soldering.

Tools and Equipment

Various hand tools for soldering.

Soldering vise.

Soldering device with dampened sponge.

Air hose.

Heat sink or antiwicking tool.

Safety glasses for eye protection.

Materials

18-gauge stranded wire with insulation (9″ long).

22-gauge stranded wire with insulation (9″ long).

Terminal board with turret terminals or an equivalent type.

22-gauge rosin-core solder.

Liquid flux and isopropyl alcohol, flux remover, or Freon vapor degreaser.

Assignment

1. Tin both ends of two pieces of 22-gauge stranded wire 4″ long (four ends tinned).
2. Tin both ends of two pieces of 18-gauge stranded wire 4″ long (four ends tinned).
3. Solder one end of each wire to a turret terminal (four solder connections).

Directions

1. Cut both wires to lengths of exactly 4″ (two wires each size).
2. Strip the insulation off from the ends of all the wires to 3/4″ (see Figure 30). Place the wire in the proper wire-size slot between the jaws of the wire stripper and carefully strip off the insulation without cutting or nicking any of the strands of the conductor. In some cases a thread or fiber tracer runs through the conductor strands. This should be removed by cutting off at the point where the insulation stops.
3. Be careful to see that the original twisted lay of the strands in the conductor is not disturbed. It may be necessary to slightly retwist the strands together.
4. Clamp the heat sink tightly on the twisted conductor up close against the end of the insulation as shown in Figure 31.
5. Use the soldering vise to hold the soldering device as shown in Figure 24. Be sure the tip of the iron is clean and coated with solder. Add flux to the bare part of the wire conductor and tin the wire with an action or movement starting close to the heat

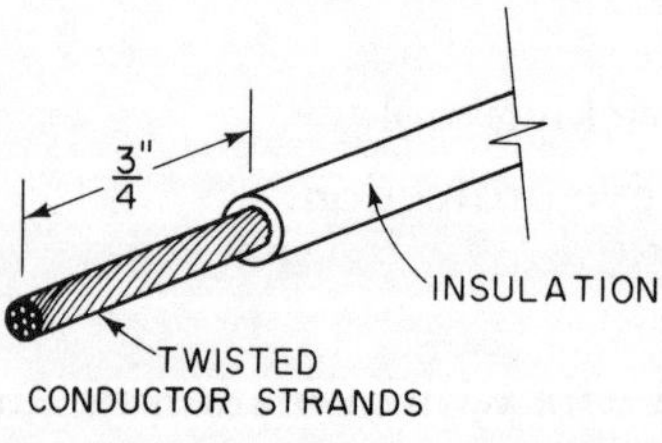

Figure 30 Insulation stripped from stranded wire

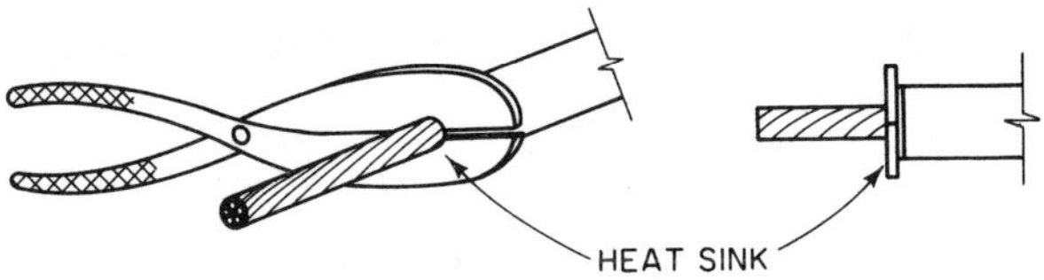

Figure 31 Heat sink clamped on wire

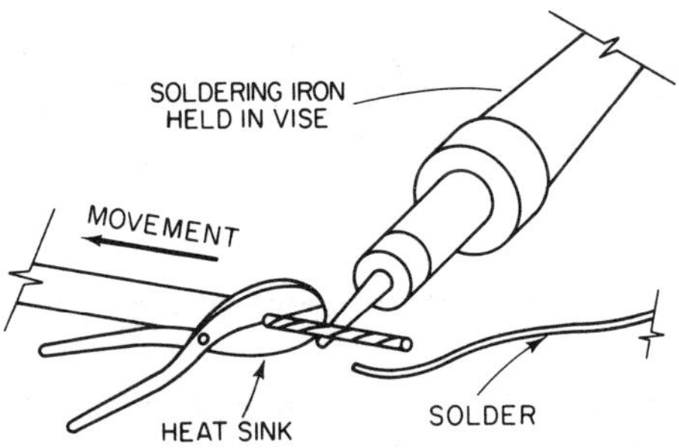

Figure 32 Tinning stranded wire

sink and moving to the end of the wire. Add solder as necessary (refer to Figure 32).

6. Prevent solder wicking by completing the tinning operation as quickly as possible and controlling each aspect of the procedure. Solder should end up at a minimum of 1 X wire O.D. from the end of the insulation (see Figure 29). Do not splash solder.
7. After the ends of each wire are tinned, connect one end of each wire to a terminal following the standards and specifications discussed in the *Introduction to Soldering* section. Note the following steps and figures of Figure 33.
   a. Place the pretinned conductor in, or against, the terminal with the insulation at a proper distance from the terminal.
   b. Wrap the conductor 3/4 turn around the terminal. Use the long-nose pliers.
   c. Cut off the excess end of the conductor, judging the length to make slightly less than one full turn.
   d. Crimp the end of the conductor in tight against terminal. Use long-nose pliers.
   e. Apply the heat sink and flux and solder the connection.
   f. Clean the terminal with alcohol or flux remover and dry with an air hose, or use Freon vapor degreaser.

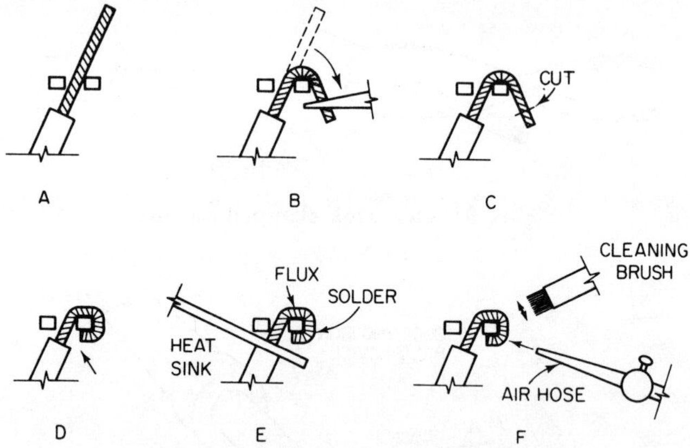

Figure 33 Making terminal connections

8. Complete the connection of one end of each wire to a terminal (two connections with 22-gauge wire, two connections with 18-gauge wire). Work safely.
9. Put your name on the terminal board and turn it in to your instructor.

COAXIAL-CABLE BREAKOUT (1)

Coaxial cable is a type of cable in which two or more conductors have a common axis. This type of cable is specially designed to operate as a transmission line at very high frequency. It is a means by which high-frequency energy is transmitted from one point to another, as in the case of an antenna to a receiver. A typical example of this type of cable is shown in Figure 34. The word "coaxial" is often abbreviated to the word "coax" and is used this way by many.

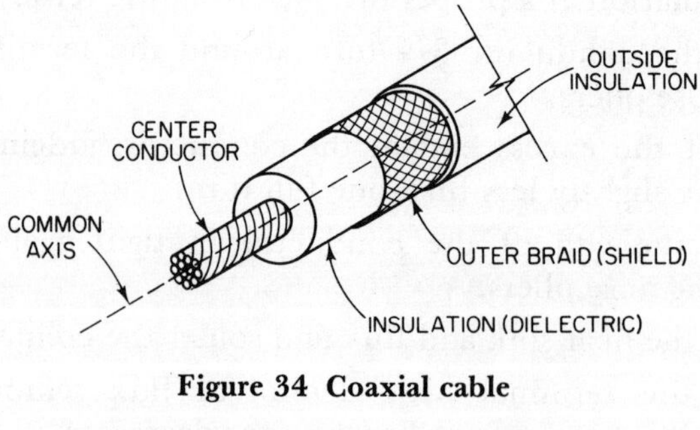

Figure 34 Coaxial cable

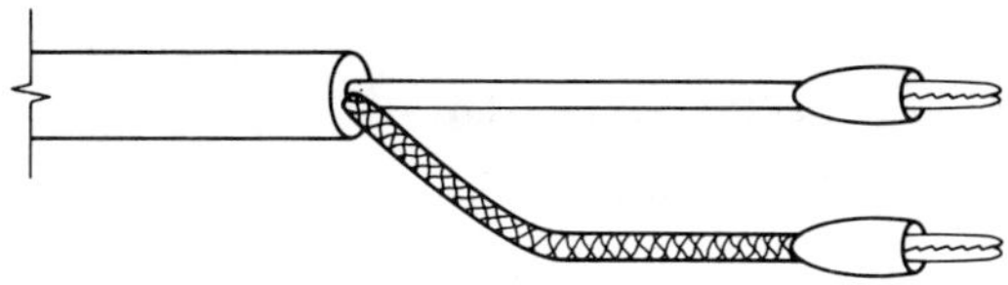

Figure 35 Coaxial-cable test lead

The physical characteristics of this type of cable are critical if it is to operate properly. These characteristics, such as diameters, spacing between conductors, and the dielectric material, determine the electrical characteristics of this type of line. The characteristic impedance (Z_0) is one important parameter that is considered when setting up a system.

For this type of line to act as it should, usually at high frequencies, the physical structure of the line cannot be disturbed. There is even a problem with connectors when this type of line is used. A change in the physical structure causes a "discontinuity," which in turn causes "reflections" of energy at that point. This, in turn, means that the energy does not end up where it is supposed to end up, as you will see when you complete further studies.

With a great deal of discretion, however, coaxial cable can be used at times when cable discontinuities are not an important factor. In these cases the center conductor is "broken out" from the outer shielding; and with the two conductors separated, the cable can be used for a special purpose of some sort (refer to Figure 35).

Purpose of Assignment

1. To become acquainted with coax and to make a coax "break-out."
2. To make a wire-braid component-lead cleaner.

Tools and Equipment

Various hand tools for soldering.

Soldering device with dampened sponge.

Soldering vise.

Scribe.

Air hose or wiping tissue.

Safety glasses for eye protection.

Materials

1′ piece of coax—RG58 or similar.

Liquid flux and isopropyl alcohol, flux remover, or Freon vapor degreaser.

22-gauge rosin-core solder, 60/40.

Assignment

1. Make a proper coax breakout 5″ from the end of the piece of coax.
2. Make a wire-braid component-lead cleaner.

Directions

1. Make the coax breakout according to the following steps. Be very careful not to nick or cut any of the wire braid which is just underneath the outer insulation.

   a. Carefully snip or tear off the outer insulation using the diagonal cutters.

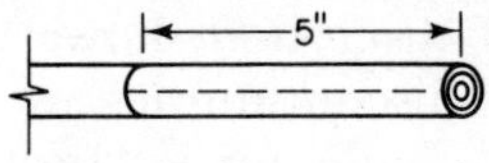

   b. Cut the end of the insulation evenly around the cable. Check for broken stands.

   c. Push the wire braid back on itself as indicated.

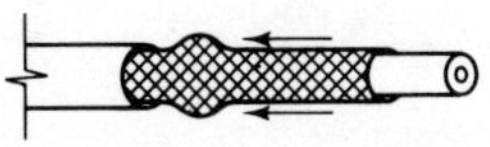

   d. Make an opening in the bulged portion of the braid using a scribe or similar tool. Slowly open the hole wider and wider.

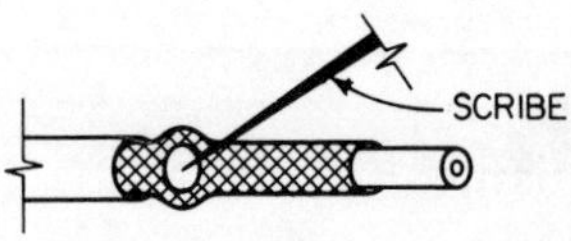

CAUTION: Exercise extreme care to avoid damage to the strands of the braid or inner insulation.

---

e. Work the scribe under the inner conductor and ease the conductor through the opening in the braid.

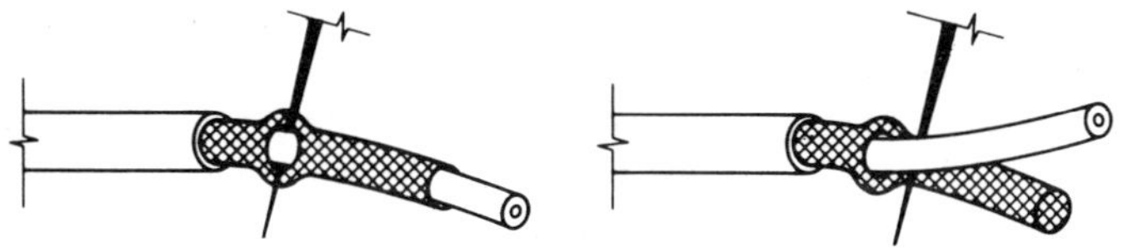

2. After the breakout has been completed and *checked,* cut off the braid close to where it breaks away from the center conductor.
3. Make a "wire-braid component-lead cleaner" as follows:

a. Push the braid inward on itself and flatten it.

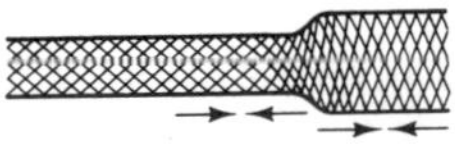

b. Solder coat the ends of the braid to 1/4″ from the ends. Work safely.

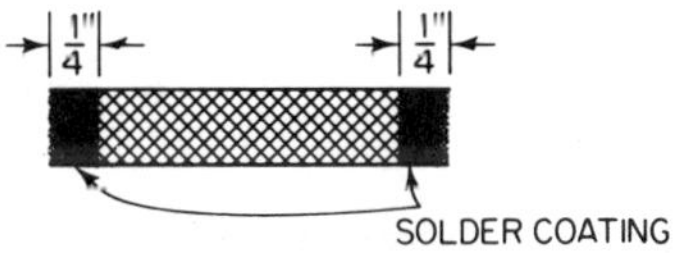

c. Trim the ends round and even with the diagonal cutters.

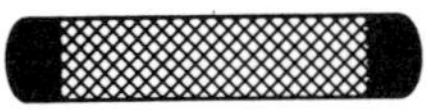

d. Fold in half for a completed component-lead cleaner.

—⁄\/\/— EST-LEAD CONSTRUCTION (1) —⁄\/\/—

Various types of test leads are usually required in order to set up and use test instruments. The circuits will vary, the test instruments will vary, and many different situations will be encountered. In some cases special test leads will be required. This means that a beginning technician should start to gather together the types of leads that are likely to be used in his future endeavors. In this assignment a few basic leads will be constructed.

Reliability is essential in order to be able to depend upon test information. All parts of the test system must be reliable, including the connecting test leads. This in turn means that the leads must be constructed carefully in order to prevent opens, shorts, or intermittents.

Purpose of Assignment

To construct test leads for use in general electronics lab.

Tools and Equipment

Various hand tools for soldering.

Soldering device with dampened sponge.

Soldering vise.

Screwdrivers—6″ and 3″.

Safety glasses for eye protection.

Materials

Kit of test lead materials as required by the instructor.

Liquid flux and isopropyl alcohol, flux remover, or Freon vapor degreaser.

22-gauge rosin-core solder.

Directions

Preliminary Information

1. Make up test leads as determined by the specific materials furnished in the test lead kit. The leads should be approximately as follows:

   a. Two 1′ red test leads (minigator clips).

   b. Two 1′ black test leads (minigator clips).

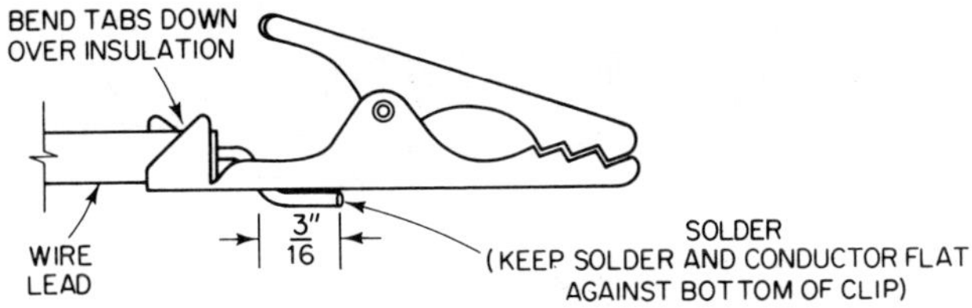

Figure 36 Minigator clip connection

c. 32″ red-black pair (single banana plugs to crocodile clips).
d. 32″ coax lead (double banana plug to crocodile clips).

2. Follow acceptable techniques and standards of workmanship in the construction of the test leads. Remember to use proper safety procedures.

Minigator Test Leads

1. Refer to Figure 36 for detail relating to the connection of the wire to the minigator clips.
2. Put a color-matching insulating boot on the wire before the lead is connected. Make the solder connection flat and smooth against the bottom of the clip and bend the tabs down over the insulation.

32″ Red-Black Pair

1. Strip the insulation off the wire for the 32″ red-black pair, 1/4″ on one end of each wire and 5/8″ on the other end of each wire. Tin all ends.
2. Connect color-matching banana plugs to the 1/4″ end of each wire. The tinned conductor is clamped down under a set screw to hold the plug connected on to the wire.
3. Put color-matching insulating boots on the wires and connect the crocodile clips on the 5/8″ end of each wire. Follow the information given in Figure 37.
4. Note especially the direction of the turn of the wire under the screw head. This is shown in Figure 38.

Coaxial Test Lead

1. For construction of the coaxial cable test lead, refer to the information given in the *Coaxial-Cable Breakout* assignment. Break out the center conductor 4″ from one end and 1¼″ on the other end.
2. Connect the double banana plug on the 1¼″ end of the coax.

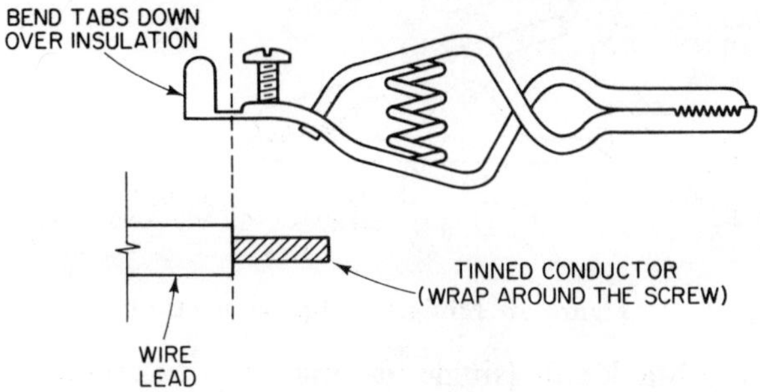

Figure 37 Crocodile clip connection

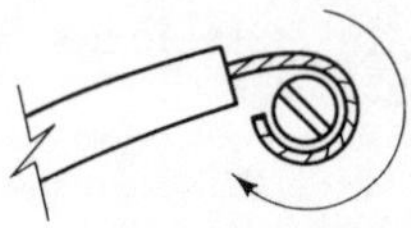

Figure 38 Connecting wire under screw

Tin the conductors and clamp them under the set screws in the plug. Note especially that the "hump" on the side of the plug indicates the ground or shield side of the coax cable.

3. Put a red insulating boot onto the center conductor and a black boot onto the braided lead on the end of the coax which has the leads broken out at 4″.
4. Connect a crocodile clip to the center conductor and also one to the braided lead. Connect the clip on the center conductor as shown in Figure 37. In connecting the clip to the braided lead, the braid can be pushed back on itself at the end, flattened, and made to fit the end of the clip. With the scribe, work a hole in the braid for the screw to go through. Tin only the part of the end of the braid that fits into the clip and under the head of the screw.
5. Have all the test leads checked by your instructor when they are finished. Slide the insulator boots back from the clips so that the connections will be easy to check.

PRINTED-CIRCUIT-BOARD SOLDERING (1)

Printed-circuit boards are made with a thin sheet of copper securely bonded to an insulation substrate. The required circuit pattern is etched into the sheet of copper by use of photographic techniques,

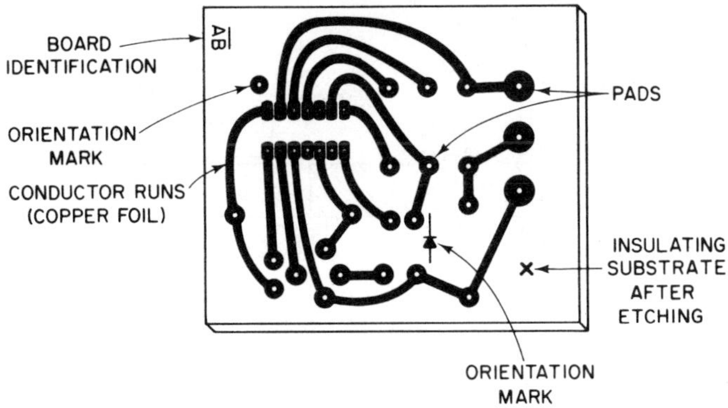

Figure 39 Typical printed-circuit board

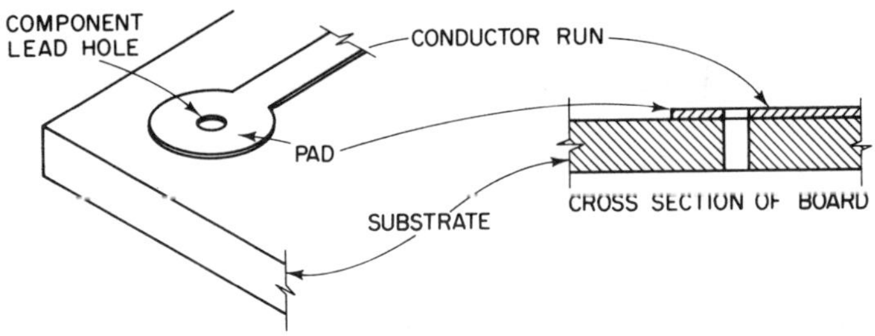

Figure 40 Typical printed-circuit board pads

resulting in a typical board as shown in Figure 39.

Enlarged views of the pad arrangement are shown in Figure 40. The views show a single-sided copper-foil type of board and include an indication of pad area and location of the component lead hole. In some cases multilayer boards are used.

In the typical arrangement of single-sided printed-circuit boards the conductor pattern will be on one side of the board (conductor side) and the components will be mounted on the opposite side (component side). The component leads are bent over and fed through holes drilled in the center of the pads to which the leads then will be soldered. Most of the time the component side turns out to be the top side of the board. Basically components can be mounted horizontally or vertically, as shown in Figure 41.

The size of the hole used for feeding component leads or terminals through the board to the conductor pattern must be carefully determined. As a standard rule the diameter of the hole should be between 0.008″ and 0.012″ larger than the diameter of the lead

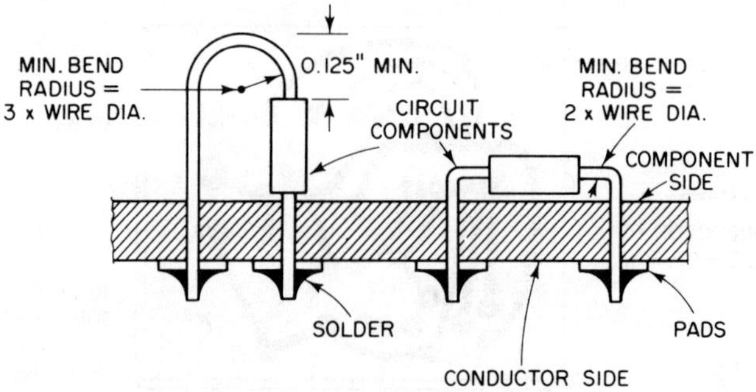

Figure 41 Components mounted on printed-circuit boards

that is to be fed through the hole. Because various component leads and terminal shanks will be of different diameters, various sizes of drills must be selected. If the lead or terminal size is not known, it can be determined easily with a micrometer or wire-gauge tool, and with the aid of a drill chart the proper-size drill can be selected. The center of the pads will usually be marked with a spot to act as a drill centering guide.

The soldering connection of the component leads to the pads usually will be done one of two ways, as follows:

1. Clip after soldering (Figure 42)
   a. Feed the component lead straight through the hole to the pad.
   b. Solder the lead with a smooth fillet, covering the pad and surrounding the lead.
   c. Clip the lead close to the top of the solder.
2. Clip before soldering (Figure 43)
   a. Feed the component lead straight through the hole to the pad.
   b. Bend the lead over 60° along the direction of the conductor run.

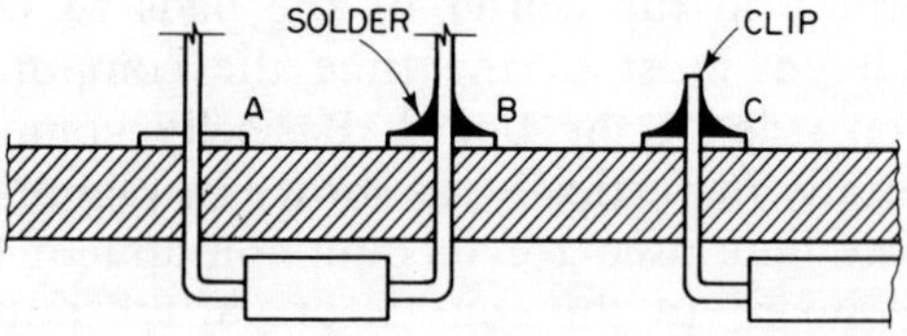

Figure 42 Printed-circuit board soldering (1)

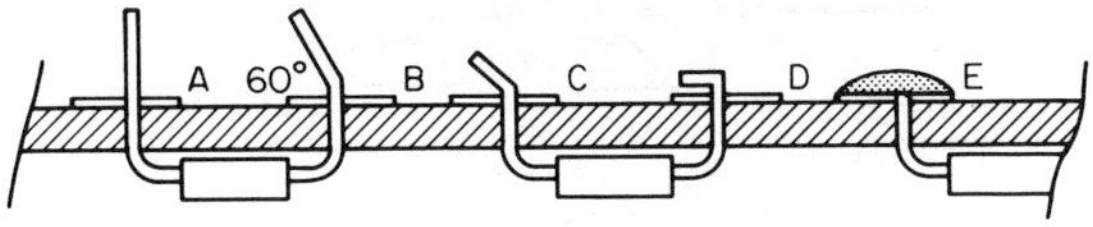

Figure 43 Printed-circuit board soldering (2)

c. Clip the lead at 1/8″.
d. Bend the lead down flat against the conductor run.
e. Solder the lead, covering the pad and forming a fillet to the lead.

Purpose of Assignment

To learn to solder to a printed-circuit pattern.

Tools and Equipment

Various hand tools for soldering.

Soldering vise.

Soldering device with dampened sponge and safety glasses.

Air hose.

Micrometer or wire gauge.

Drill-size chart and various drills from about #54 to #70.

Materials

Printed-circuit board ready for component mounting (this can be a reject board from industry).

Miscellaneous components for soldering practice, if available.

18-gauge jumper wire.

22-gauge rosin-core solder.

Liquid flux and isopropyl alcohol, flux remover, or Freon vapor degreaser.

Assignment

Obtain an etched printed-circuit board ready to receive components, and practice soldering components or bare-wire jumpers to the board.

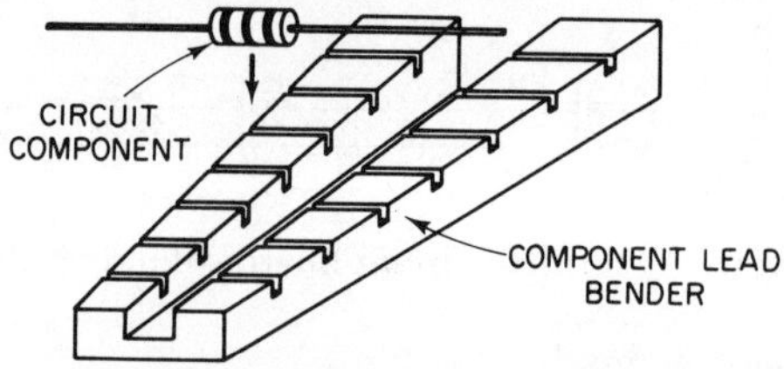

Figure 44 Bending component leads

Directions

1. Obtain a printed-circuit board that is ready to receive components. This can be a reject board from a production line or a board made up for soldering practice.
2. If it is necessary, drill holes in the board for receiving the component leads or bare-wire jumpers. Refer to the information given earlier in this section relating to hole size and drill selection.
3. Bend the component leads over to match the pads using a component lead bender, or bend the leads by holding them at the proper place with long-nose pliers (refer to Figure 44).
4. Mount components and/or bare wire jumpers for soldering practice. Try both methods of soldering suggested in the introduction to this section. Be sure to add flux at each connection.
5. Clean the board of all flux residue after soldering, and protect your eyes.
6. Inspect the connections for the proper amount of solder and uniformity.
7. Put your name on the board and turn it in to your instructor.

DE-SOLDERING TECHNIQUES (1)

Solder follows heat. This is a good rule to know, for it is the basis of the operation of certain de-soldering techniques. In these techniques the solder is removed from a connection by making the solder flow towards the heat, which is applied to a somewhat porous wire braid or stranded wire. The braid or wire melts the solder, which, in turn, absorbs the solder. This action works so smoothly that with little effort practically all traces of solder can be removed from a connection. Figure 45 shows this action.

Small spools of wire braid are available commercially. This braid

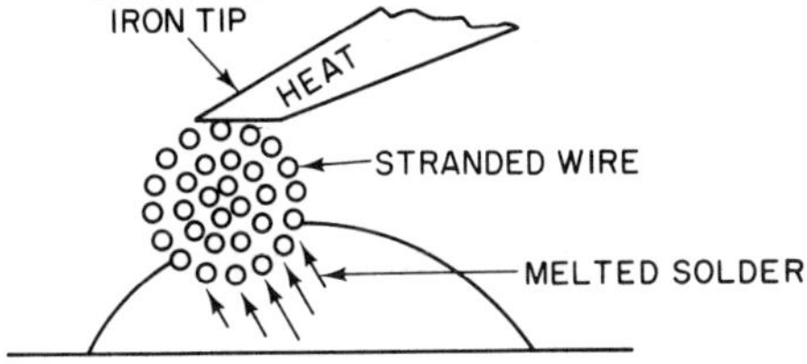

Figure 45 Solder being absorbed by heated wire

is made of a very fine tin-coated wire and is usually also coated with a noncorrosive, electrically nonconductive rosin flux. The use of this braid is very satisfactory. It has also been determined, however, that tremendous results can be obtained by using a flux-coated stranded wire for this operation. An 18-gauge stranded wire works well for removing fairly large quantities of solder and a fine-stranded 22-gauge wire is excellent for removing small quantities of solder or the last traces of solder from a connection. The use of stranded wire will be discussed in this assignment.

Figure 46 shows before and after views of the solder-removing action. Printed-circuit pads are used in this illustration, but the technique applies equally well to other types of connections.

This is a fast, clean, and easy-to-use technique. No special equipment is required other than ordinary soldering tools, including a regular 30- to 50-W iron. One of the chief advantages of this technique is that during the procedure very little heat is transferred to the connection, which is a very distinct advantage for the protection of heat-sensitive parts.

In using the stranded wire for this solder-removing procedure the wire must first be prepared prior to its use. This is done by stripping the insulation back to a distance of approximately 1″, bending

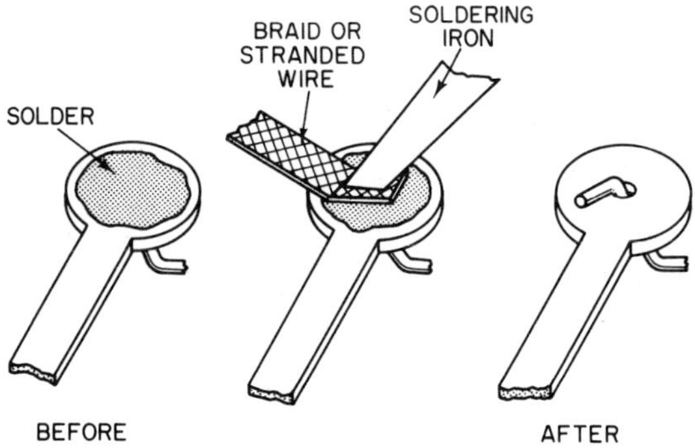

Figure 46 De-soldering a connection

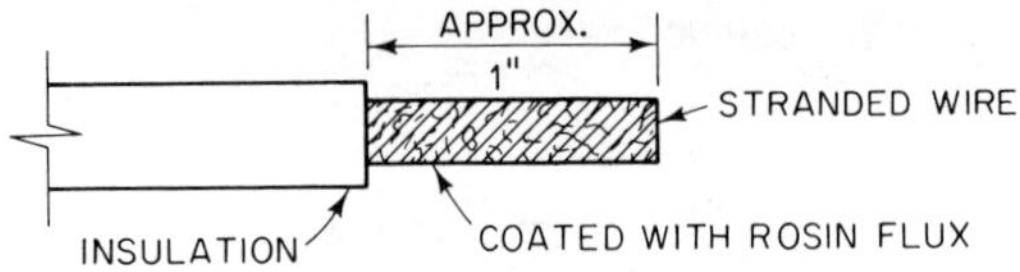

Figure 47 Wire prepared for de-soldering

the wire slightly, and adding flux to the bare end of the wire. The wire is then ready for use, as shown in Figure 47.

After the wire is prepared for use it is placed on top of the solder to be removed and the soldering iron is placed on top of the prepared wire. Heat is transferred through the prepared wire to melt the solder and the solder is absorbed up into the wire. Figure 48 shows this arrangement.

It should be stated that there are various vacuum suction devices that are also used for de-soldering. These work fairly well and in some cases are needed when the solder is far down in a hole. The main disadvantage of the use of most of these devices is that the connection itself must be heated to melt all the solder. This necessitates the use of considerable heat on the connection as compared to the method described above.

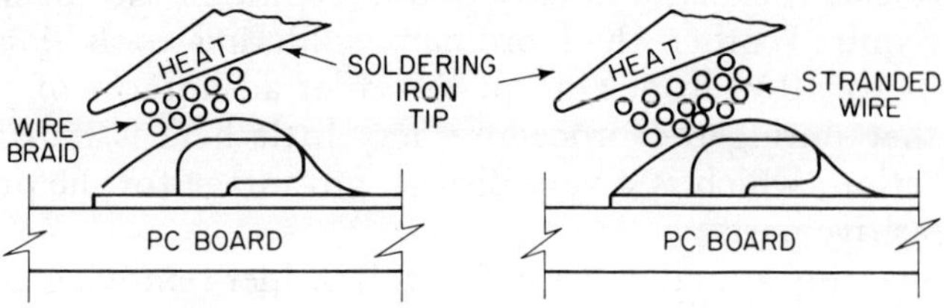

Figure 48 Removing solder from a connection

Purpose of Assignment

To learn the technique of using stranded wire for removing solder from a connection.

Tools and Equipment

Various hand tools for soldering.

Soldering vise.

Soldering device with dampened sponge and safety glasses.

Air hose or wiping tissue.

Materials

18-gauge stranded wire.

22-gauge stranded wire.

Printed-circuit board with completed solder connections.

Liquid flux and isopropyl alcohol, flux remover, or Freon vapor degreaser.

Assignment

Remove the solder from several connections on the printed-circuit board. (If jumpers or components are soldered to the board, remove the solder from only one end of each jumper or component and leave the end of wire in the hole from which the solder was removed.)

Directions

1. Prepare, as needed, 18- and 22-gauge stranded wire as described in the introduction to this section.
2. Lay the prepared end of the stranded wire on top of the solder to be removed.
3. Apply the tip of the soldering iron to the top side of the stranded wire and allow the wire to absorb the solder. Refer to Figure 48 and repeat the operation as many times as needed to remove all the solder. Try both sizes of stranded wire. Tin the soldering iron so that there will be a *solder bridge* and a quick transfer of heat from the tip of the iron to the stranded wire, which in turn will melt the solder.
4. After the solder is removed, clean the connections of flux residue using alcohol, flux remover, or Freon vapor degreaser. Blow the board dry with the air hose, or wipe dry with a wiping rag (unless Freon vapor degreaser is used). Remember eye-protection procedures.
5. Put your name on the board and turn it in to your instructor.

PRINTED-CIRCUIT-BOARD CARRIER (1)

Very often in the manufacturing process, a jig, a device, or a tool of some sort is used as an aid in performing certain operations. In many cases this device is specially designed and constructed to overcome certain production-line problems.

The printed-circuit-board carrier is such a device as mentioned above. It has been designed as a carrier for small printed-circuit boards during certain steps involved in their fabrication. With this device the board is held at only the bottom edge of the board, thus alleviating problems that would occur if the board were held at its top edge. With the top edge free and clear, the various solutions will more easily drain down the sides of the board without causing build-up or unevenness. This is especially important in the case of coating the board with KPR using the dipping method.

It should be noted that the carrier should be made to fit the board correctly so that there is not the danger of losing the board in the tanks containing various solutions.

Purpose of Assignment

1. To construct a carrier for holding printed-circuit boards during fabrication operations.
2. To develop skill in observing pertinent detail in the construction of an item.

Tools and Equipment

Lineman's pliers or equivalent.

Small cleaning brush.

Materials

Ordinary wire coat hanger with a wire diameter of approximately 0.090 to 0.098″.

Paint remover.

Water-rinse facilities.

Assignment

Using the wire from a coat hanger, make a printed-circuit-board carrier. Refer to the information in the Directions.

Directions

1. Using the lineman's pliers, cut the hook off the coat hanger, leaving approximately 24 to 30″ of wire.
2. Straighten the wire into a straight strand.
3. Remove the paint or varnish from the wire using a small amount of paint remover. Water-rinse the brush and wire after the paint has been softened and wrinkled or removed from the wire.

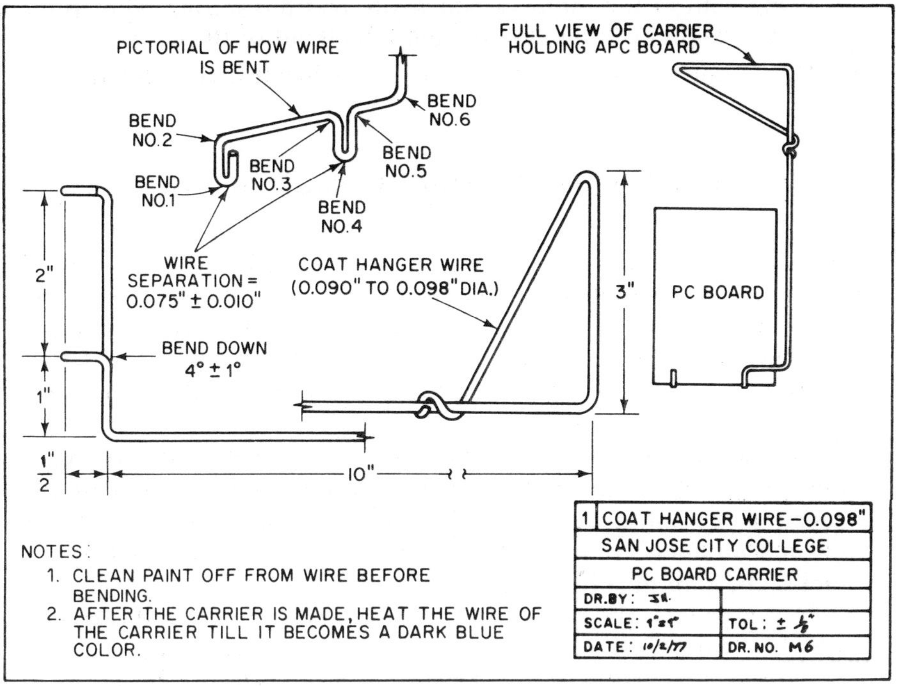

Figure 49 Printed-circuit-board carrier

4. Bend the wire to make the carrier according to the directions given in Figure 49. Make the bends in the order given. Also, make the bends at the bottom of the carrier according to the dimensions given in the drawing. The bends for the handle part of the carrier can be made by making the wire fit the pattern porportions given in the drawing.
5. The small portions of wire forming the two loops at the bottom of the carrier should have a gap that is 0.075 ± 0.010″, and the wires should be parallel at those points.
6. Note that when the carrier is finished, it should be heated until it becomes dark blue in color. This forms an oxide coating on the wire, and tin from the electroless tin-plating solution is prevented from coming out of solution and adhering to the wire of the carrier. The tin should go to the copper on the printed-circuit board, not the carrier wire.
7. Also note especially that there is a 4° bend made at the position of the inside loop on the bottom part of the carrier. *This bend is the secret by which the board is held in the frame.* As a result of the bend, there is a slight spring pressure on the board that is

being held in the loops in which the board is setting. This spring pressure will keep the board from falling out of the carrier.

8. Have your instructor check the carrier when it is completed.

PRINTED-CIRCUIT-BOARD ETCHING (KPR) (1)

The etching process is one of the processes used in industry in the fabrication of printed-circuit boards. In this process very fine detail and good line definition can be obtained on a mass production basis, and it is a process that is adaptable to many variations. Light-sensitive emulsions, silkscreen printed patterns, or plated patterns are used in various steps of the process.

This assignment will deal with the use of a light-sensitive emulsion for obtaining the pattern which will resist the action of the etchant in certain areas on the board. Photographic techniques are used in conjunction with this emulsion to achieve high quality and exacting pattern detail.

In this process the copper side of a printed-circuit board is coated with the light-sensitive emulsion (Kodak Photo Resist), the emulsion is exposed through a film, developed, and then the background areas, which are not protected from the etchant, are dissolved in the etching solution. Refer to Figure 50, which illustrates the different steps of the sequence.

The use of the emulsion is dependent upon *polymerization.* In this process the portion of the emulsion that has been exposed to light is physically restructured by the union of molecules, which in turn causes a new compound to be formed. This new compound is insensitive to the developer. In the unexposed areas, the emulsion will be dissolved by the developer.

The emulsion coating is not sensitive to light while it is wet. After the coating is dry, care must be taken to handle the coated board in a subdued light.

The copper surface to be coated should be a smooth, clean surface so that pinholes will not form in the emulsion. The KPR is like a lacquer and will adhere to smooth, acidified surfaces which are obtained by dipping the board in a bright-dip acidifier. After development, the emulsion is soft and should not be touched until after it is rinsed with cold water and dried in an oven. The cold-water rinse and oven bake will rinse and remove solvent residues, and harden the emulsion coating.

For the purpose of this assignment, it is to be considered that

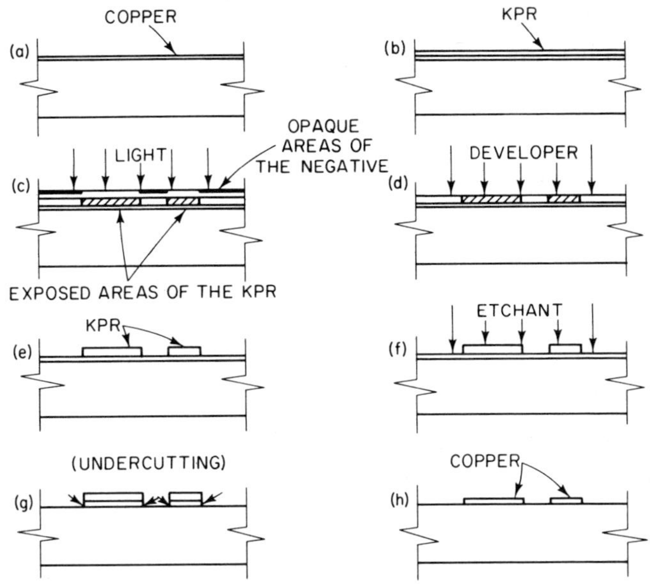

(a) Basic PC board
(b) KPR coated board
(c) Coated board exposed through the negative
(d) The pattern is developed
(e) Developed pattern after background is clean
(f) Exposed copper is etched
(g) Undercutting at the edges of the pattern
(h) KPR is removed and the copper is tin coated

Figure 50 Printed-circuit-board etching sequence

the artwork for the circuit has been completed and that a photographic film negative has been obtained. Refer to *Spiral-Sweep-Generator Artwork*(1) and *Printed-Circuit Artwork Development*(1) for information on this.

Purpose of Assignment

1. To learn printed-circuit-board etching techniques.
2. To gain experience in the use of photosensitive resist materials.

Tools and Equipment

Solution trays or containers.

Light source for exposure (see Directions).

Bake oven.

Semidark room with 10-W amber light.

Contact printing frame.

Compressed air.

Solution-heating facilities.

Water-rinse facilities.

Freon vapor degreaser.

Goggles for eye protection.

Materials

Film negative of circuit layout pattern ($\overline{AB}$ circuit supplied by the instructor).

Printed-circuit-board stock.

Detergent.

Hydrochloric acid (approximately 20%).

Kodak Photo Resist (KPR3).

Ortho Resist Thinner.

Ortho Resist Developer.

Ferric chloride etchant (40° Baumé solution).

Water.

Chemical immersion tinning process.

Plastic-monofilament line—18″.

Assignment

Obtain the negative for a printed-circuit pattern. Etch the pattern on a printed-circuit board following the instructions given in the Directions. This can be the circuit pattern developed in *Spiral-Sweep-Generator Artwork*(1), *Printed-Circuit Artwork Development*(1), or any other circuit, such as one assigned by the instructor.

Directions

Work carefully using the following directions. Keep the board and work areas free from dust, which will impair the uniformity of the coating. Also, keep the film *protected* from being scratched and from being wet by the solutions. Use the printed-circuit-board carrier which was made earlier.

1. Prepare the copper surface.

   a. Degrease the surface by scrubbing with a fine buffing powder such as pumice, or by using a detergent solution and the Freon vapor degreaser. Water-rinse thoroughly.

b. Dip the board in a 20% solution of hydrochloric acid to remove oxides and dirt and to acidify the surface. Thoroughly rinse the board in water.

c. Dry the board quickly without letting drops of water stand on the surface, and do not touch the surface with hands after it is dry. The Freon degreaser can be used for drying the board.

NOTE: Steel wool is sometimes used in preparation of the surface, but it is very dirty and leaves steel-wool dust, which can be a detriment in subsequent operations. Steel wool also leaves a rough surface, which is detrimental to smooth resist coatings.

---

2. Coat the copper surface with the KPR emulsion, using any effective method. A spray, brush, dip, roller, or whirler can be used. Work under a subdued light (10-W amber bulb). Make the coating thin and as even and smooth as possible. Also keep it from being thick in spots. Initially, the KPR3 should be diluted using 1 part KPR to 1 part thinner. If the board is dipped to coat it, use the printed-circuit-board carrier to hold the board. The carrier holds the board at the bottom edge of the board and will aid in getting smooth, undisturbed coatings.
3. Air-dry the board for 1 to 2 minutes and then prebake 10 to 60 minutes at 140 to 200°F.
4. Mount the coated board and film in a contact printing frame. If a commercial vacuum printing frame is not used, place a clear piece of Plexiglas over the film to hold it *tight* against the coated board. This is important. *All parts* of the film must be held tight against the coated board. Refer to Figure 51 and be sure that the film is facing in the right direction.
5. Expose the coated board through the film, using a proper light source. A carbon-arc lamp, sunlight, or in some cases a photoflood light may be used. Expose for 3 minutes if direct sunlight is used. For other light sources, determine the exposure time by experimentation.
6. Develop the image for 2 or 3 minutes in a tray using KRP developer, with considerable agitation.

   a. The image can be dyed at this point if so desired. Dip in KPR dye for 30 seconds but do not dye the emulsion unless it is necessary to improve the coating for inspection or retouching.

   b. Water-rinse the board with *cold* water. Rinse until the emul-

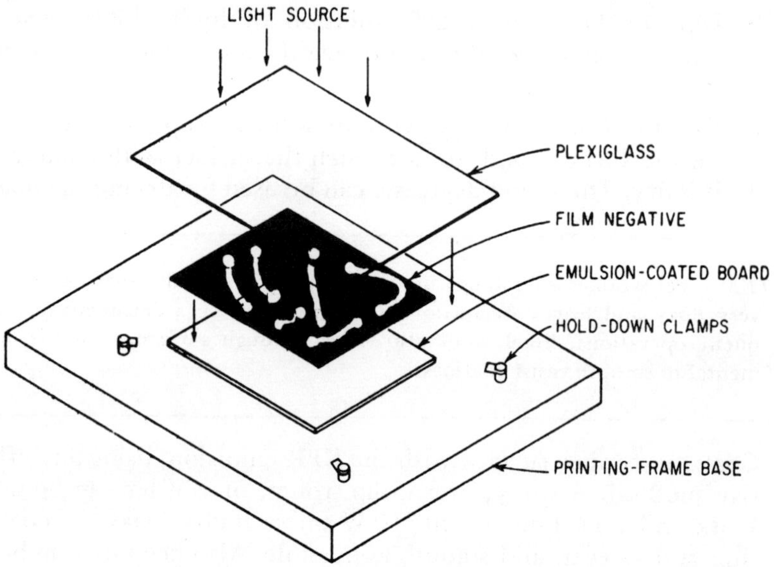

Figure 51 Exposure-frame arrangement

sion is hardened by the cold water and solvent residues are rinsed away.

c. Look for the effects of underexposure or overexposure. The background is not clear if *overexposed* and the pattern is not clear if *underexposed.*

d. Postbake, if desired, for 10 to 15 minutes at 300 to 500°F to remove solvent residues and improve resist qualities.

7. Etch the pattern using a 40° Baumé solution of ferric chloride. Refer to Appendix A for information on the measurement of the solution density. Chromatic acid or ammonium persulfate also can be used as the etchant if desired.

   a. Heat the solution to about 120°F.

   b. As a check on the pattern, if the emulsion has not been dyed, quickly dip the board in and out of the etchant to make the pattern stand out. Inspect the pattern carefully, and if necessary, the pattern can be retouched or the process can be repeated from step 1.

   c. Etch for approximately 5 to 10 minutes or until the background is dissolved. Use airflow agitation and be sure that the air is free of oil. Refer to Figure 52, Figure 187, and Table A-5. In some cases the board can be etched in an open tray

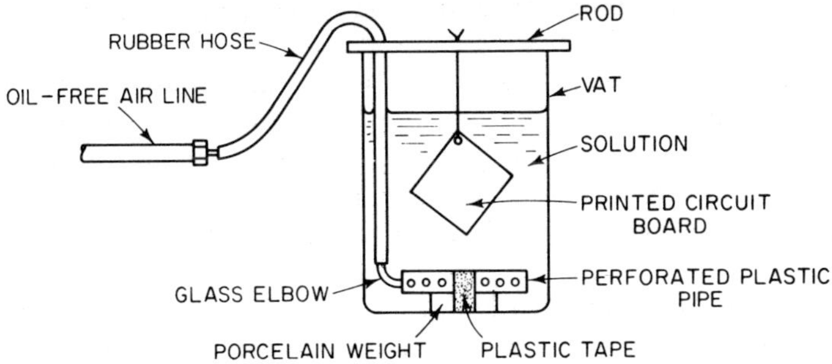

Figure 52 Etching-tank detail

that is rocked back and forth so that the ferric chloride will continuously wash across the board.

d. When the background is completely etched, water-rinse and dry the board. Do not etch the board to the extent that there will be excessive undercutting, which is damaging to the pattern detail. *Use eye-protection.*

8. Remove the resist image by using paint remover or KPR stripper. Hold the board 2 minutes in the stripper, water-rinse, scrub with a nylon pad, and degrease.
9. After the board is thoroughly clean, tin-coat the board using an electroless chemical immersion process. Dip the board for 2 seconds and then check for adequate coverage. *Use eye-protection and do not splash solutions.*
10. If coverage is adequate, dip the board for 5 to 10 minutes with slight agitation. Have your instructor check the board before continuing. When the above steps have been completed, the board is ready for subsequent fabrication operations. These operations are described in *Mounting Printed-Circuit Components.*

MOUNTING PRINTED-CIRCUIT COMPONENTS (1)

(continued from previous assignment)

Circuit reliability is a factor foremost in the mind of the designer from the first calculations of the parts of a circuit to the final stamp of approval on the constructed and tested product. Careful and precise workmanship is therefore an absolute necessity, even though techniques and sequence steps may vary. Basic information on board construction is given in this assignment.

The orientation of components on a board is important. Boards are usually designed so that the components are parallel to each other and parallel to the edges of the board. In addition to this, the person assembling the components on the board is responsible for the proper orientation of parts. Color bands should face in the same direction for ease of reading, and the wording of values or labels should be made readable. The parts should be square, or straight and level on the board. In some cases the components can be mounted on edge or on end. Refer to the introduction of the *Printed-Circuit-Board Soldering* assignment.

A minimum radius of 2 X wire diameter should be used in bending the component leads, and in some cases the leads can loop back under the body of the component before bending down to go through the holes in the board (see Figure 53). The body of the component should be centered between where the leads bend down. Using a component-lead bender makes it easy to accomplish this.

The information on the placement or the location of the components on a board is usually given in a *component mounting schedule* drawing. The circuit pattern is given as seen from the component side of the board and will include the use of component symbols and part numbers to show where the components are placed in the pattern. Refer to Figure 54, which shows a typical component mounting schedule.

After it is known where the components will be placed in the circuit pattern, a *drill schedule* is made up. This schedule will show the location and size of the holes to be drilled on the board. On this schedule a table will show various sizes of drills matched with a letter to represent each size. The letters will appear on a drawing of the circuit pattern to show where each size of hole will be drilled. Refer to Figure 55, which shows a typical drill schedule.

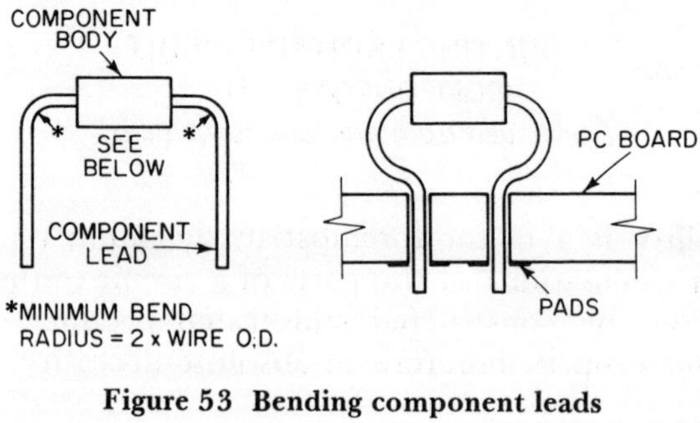

Figure 53 Bending component leads

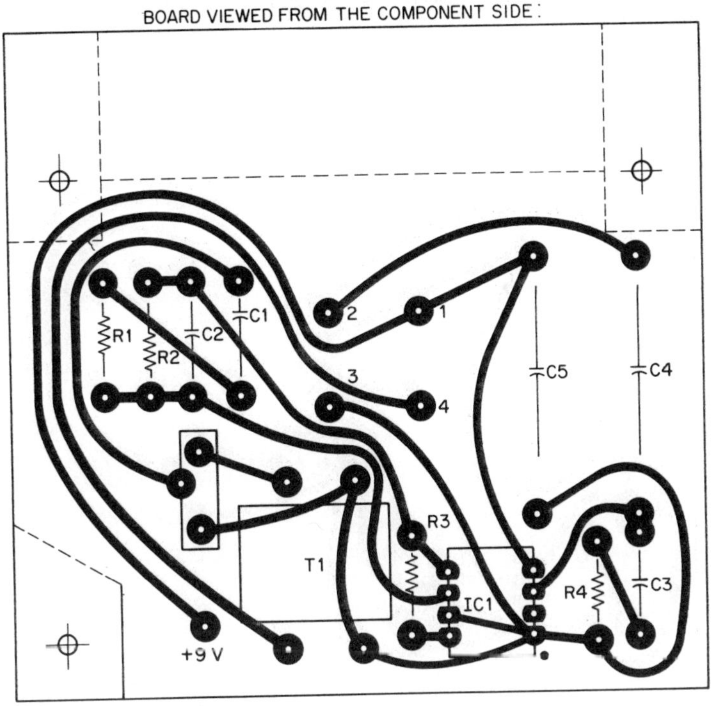

COMPONENT MOUNTING SCHEDULE	
R1	10 kΩ $\frac{1}{2}$ W
R2	1 MΩ $\frac{1}{4}$ W
R3	1 kΩ $\frac{1}{2}$ W
R4	1 Ω $\frac{1}{2}$ W
C1	0.05 μf
C2	51 pf
C3	0.05 μf
C4	200 μf, 6 V
C5	100 μf, 12 V
IC1	MC 1306 P
T1	D1−712

Figure 54 Typical printed-circuit board component mounting schedule

Holes are drilled in the center of the pads of the circuit pattern and through the substrate for the purpose of allowing the pigtail leads of the components to feed through and make connections to the circuit pattern. The size of these holes should be closely controlled so that the solder can more easily fill the gap between the lead and the pad. It is recommended that the diameter of the

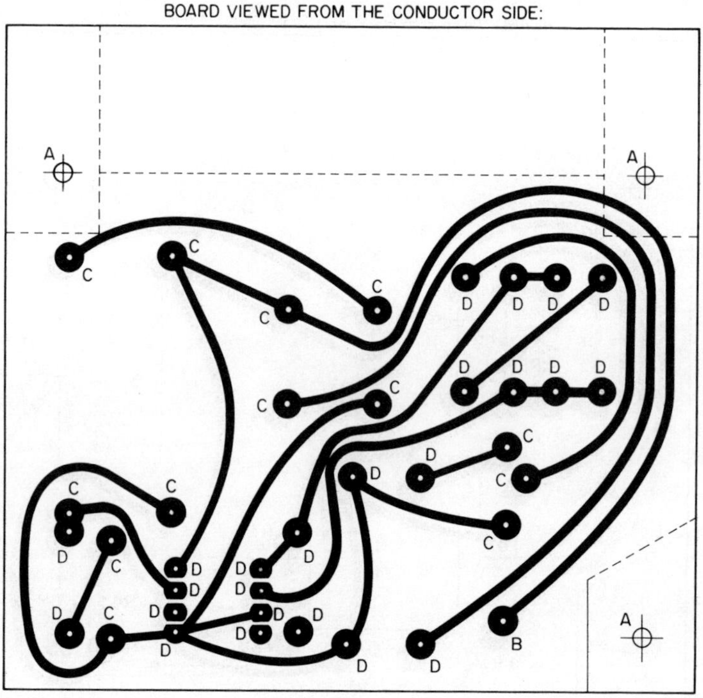

DRILL SCHEDULE	
A	# 28 DRILL
B	# 42 DRILL
C	# 60 DRILL
D	# 70 DRILL

Figure 55 Typical printed-circuit-board drill schedule

hole be somewhere between 0.008 and 0.012″ greater than the lead that is to go through the hole. This is the normal type of specification that would apply to the mounting of ordinary components. For certain items, such as turret terminals, the hole size should be within 0.002 or 0.003″ greater than the part that goes into the hole. In these cases it is necessary that there be a tighter fit so that physical strains on the part would not be entirely taken up by the solder. The information for the location of holes should be given in the *hole layout drawing.* The location of the components in the layout is found in the *component mounting schedule* (see Figure 54), and the information for the size of the holes should be given in a *drill schedule* as shown in Figure 55.

Tinning the conductor pattern before the components are

mounted is often done in order to cover the conductors with a protective coating. It is also done because it is an easy way to make reliable solder connections. Experience shows that there is merit in both of these points. This is usually not done, however, if the pattern has been plated. On the contact fingers, this is especially true where a gold plating has been applied for protection of the spring contact surface. It is suggested that in the small shop where facilities are limited, a *thin coating* of solder on the contact surfaces is better than bare copper, which allows oxide to form. In tinning the conductor pattern, the copper must be *perfectly clean* when using a chemical immersion process for tin coating.

Proper quantities of solder should be used at each connection. Usually, a smooth fillet of solder is correct when applied with the proper amount of heat. Refer to other assignments for information on this. It should also be noted that soldered boards should be properly cleaned in preparation for subsequent operations.

Turret terminals and eyelets are often mounted on printed-circuit boards. Terminals are mounted for the purpose of making external connections to the board by the soldering of jumper wires or components to the terminals. This is shown in Figure 56.

Eyelets are used where two or more pigtails will connect to the board, or they are used to act as a feed-through on double-sided boards, or to keep the pads from lifting from the board. Their use is often simply a matter of preference. For detail on the mounting of these items, refer to Figure 57.

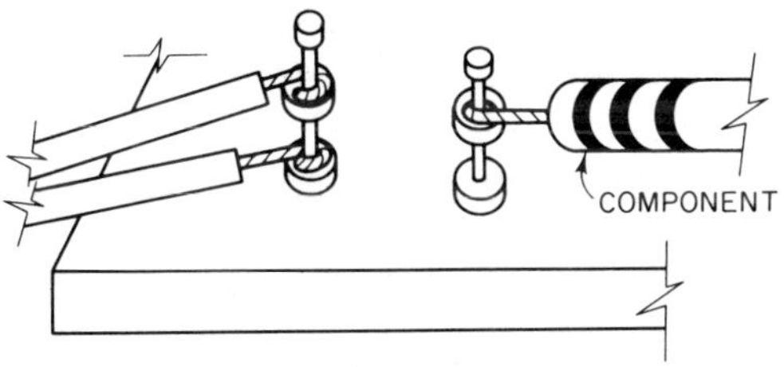

Figure 56 Turret terminal connections

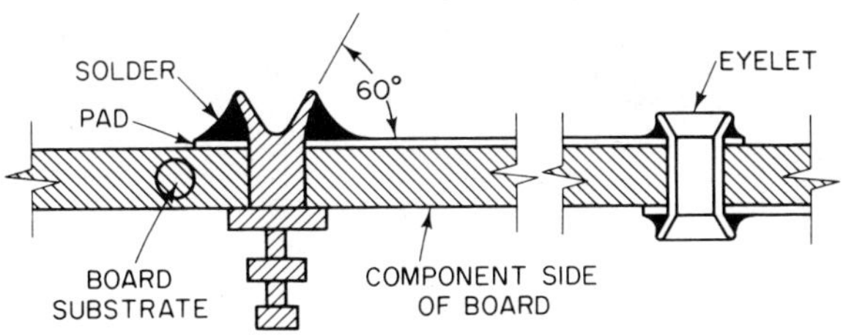

Figure 57 Turret terminal and eyelet soldering

The funnel type of expansion is superior to the flat swaged-over arrangement. It allows the solder to flow more easily between the edges of the shank and the pad. The chances for a good connection are much better.

To mount a turret terminal, the shank is flared to 60° only, deliberately not down flat against the pad. This is to allow the solder to easily fill the space between the flared shank and the pad. Refer to Figure 57 for an illustration of this. The inside of the hole of the shank does not need to have solder in it.

A base for mounting turret terminals can be just a heavy piece of steel with the proper-size hole drilled to fit the top post of the terminal. The hole cannot be so large that it will not allow the base of the terminal to seat against the steel during the mounting operation. Figure 58 shows a cross section of how the terminal should fit in the steel base and how the 60° punch is used to expand the shank of the terminal.

Parts identification labeling is usually applied to the board just prior to mounting the parts in position. These labels should be on the front side (component side) of the board and located just off to the side of where the actual parts are positioned. Rubber stamps with ink or paint can be used for the application of the labels, or the screen-process stencil can be used where the pattern must be repeated. Refer to the introduction of *Printed-Circuit Artwork Development* (1) for further information.

A sprayed-on protective coating is often applied after the components are mounted and soldered and the board is cleaned. The use of this coating, of course, depends on individual requirements, and also the type of coating may vary. Usually, good environmental protection can be obtained with varnish, lacquer, acrylics, or epoxies. Extreme conditions will require the application of special materials.

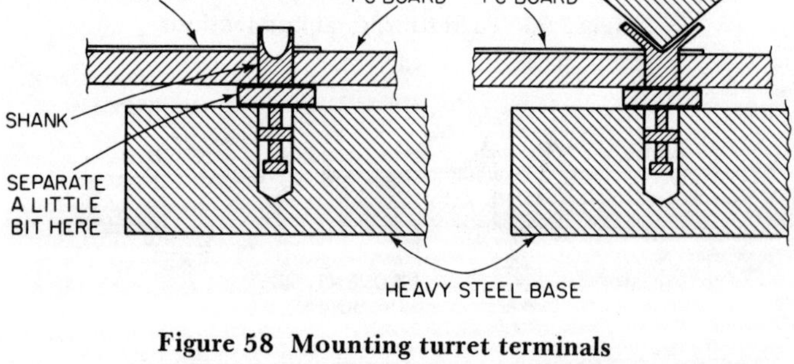

Figure 58 Mounting turret terminals

Purpose of Assignment

1. To learn the techniques of mounting components on a printed-circuit board.
2. To gain experience in the accomplishment of neat and careful workmanship and safety procedures.

Tools and Equipment

Various hand tools for soldering.

Soldering vise.

Soldering device with dampened sponge.

Air hose (used with flux remover).

Micrometer or wire gauge.

Drill-size chart.

Drill press.

60° punch.

Squaring shear.

Goggles for safety.

Materials

Printed-circuit board ready for component mounting.

Components assigned for printed circuit board.

Turret terminals.

22-gauge rosin-core solder.

Liquid flux and isopropyl alcohol, flux remover, or Freon vapor degreaser.

Assignment

Mount the components on a printed-circuit board designed and prepared for a particular circuit. This can be the circuit made in the *Printed-Circuit-Board Etching (KPR)* assignment.

Directions

It is assumed at this point that the artwork for the board has been completed, and the board etched and tinned, leaving the board ready for subsequent operations. For laying out the board, etching, and tinning the board, refer to assignments relating to circuit-board layout and etching.

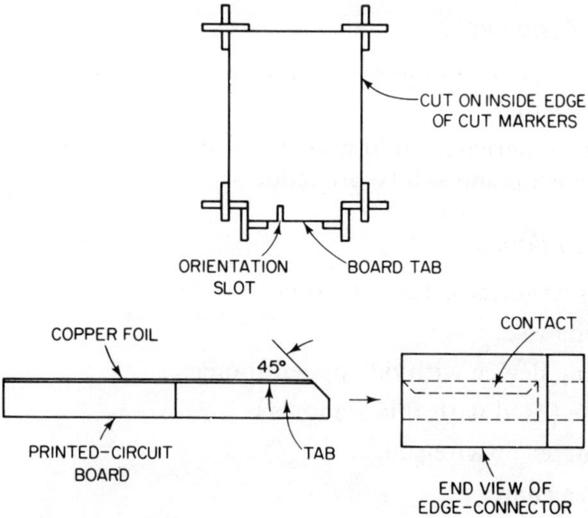

Figure 59 Beveled edge of edge-connector tab

1. Trim the board on the inside edge of the cut lines and carefully smooth up the edges of the board with a flat, smooth file. Trim the board to size using the squaring shear for straight cuts or the bandsaw for cuts into a corner such as on a board tab. If applicable, bevel the edge of the tab which fits into the edge connector. Clean up the edges of the board with a file (refer to Figure 59), and file a 45° angle on the tab part of the board that slips into the edge connector.
2. Select the proper components for the board or use components that have been assigned by the instructor.
3. If a component mounting schedule is not furnished, trace the circuit layout on paper and make up a proper *component mounting schedule.* Refer to Figure 54 for information. If the $\overline{AB}$ circuit layout is used, fill out the information in Figure 60 to make a complete and proper component mounting schedule.
4. Fill out the information in the table for the selection of drills. Use a micrometer (see Appendix B) to measure the component leads in *mils.* Make the proper calculations and use the drill charts for drill-number information. When a choice of drills is made for each component, make a final selection involving the smallest number of drills possible that will meet the requirements for all the holes.
5. Using the information obtained in step 4, make a complete drill chart and drill schedule (refer to Figures 55 and 61).

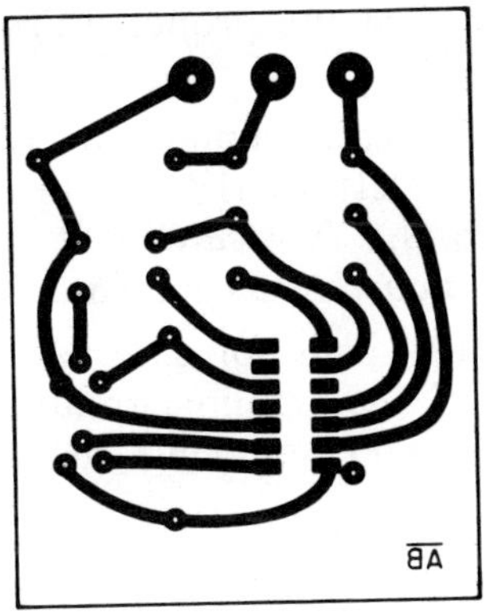

COMPONENT MOUNTING SCHEDULE	
PART NO.	VALUE

DRILL SELECTION CHART

COMPONENT NO.	LEAD DIAMETER	HOLE SIZE ALLOWED		DRILLS ALLOWED		FINAL DRILL SELECTION
		MIN.	MAX.	SMALLEST	LARGEST	

Figure 60 $\overline{AB}$ component mounting schedule

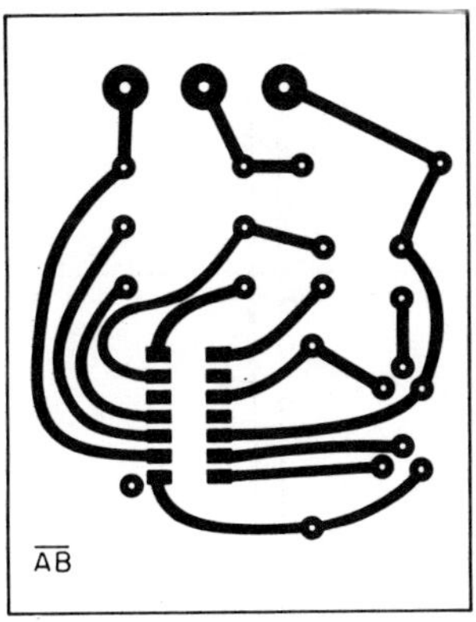

DRILL SCHEDULE	
A	DRILL #
B	#
C	#
D	#
E	#

Figure 61 $\overline{AB}$ layout drill schedule

6. Drill the component-lead holes. Select the proper drill size for each hole. Wear eye protection.
7. If applicable, label the board for each component with the proper part number, reading from left to right or bottom to top as per instructions. Refer to the *Printed-Circuit Artwork Development*(1) or *Screen-Process Labeling* assignments.
8. Mount turret terminals in the large pads for external connections to the board. The shank of the terminal should be expanded in the board hole to approximately a 60° angle. Refer to the introduction for information.
9. Bend the component leads to fit the distance between mounting holes. Make 90° bends with a minimum radius approximately equal to 2 X wire O.D.

   a. The components should be mounted so that their value is readable.

   b. Install components with color bands to read in the same direction.
10. The integrated circuit must be properly oriented on the board. Pin 1 of the integrated circuit must face the reference dot on the pattern, or in some cases the pad for pin 1 is rectangular in shape. Do not clip the integrated-circuit leads that extend through the board.
11. Mount the transistors 1/8″ off the board and be sure they are not slanted or tipped.
12. Carefully solder all connections. Be sure that there are no holes

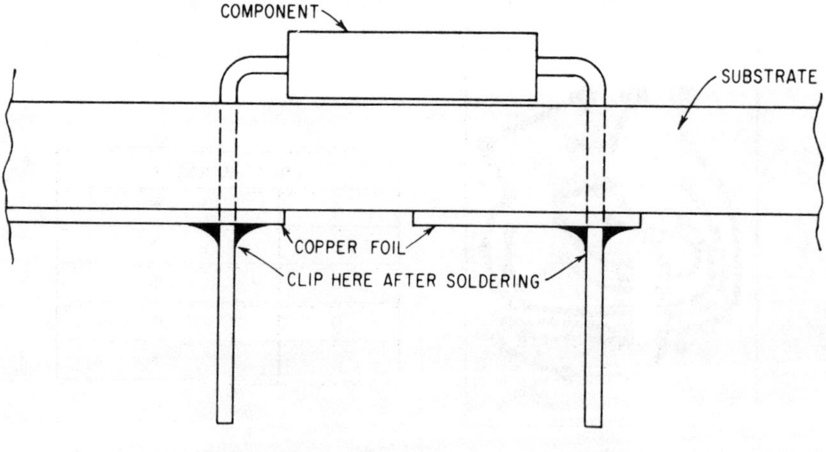

Figure 62 Leads clipped after soldering

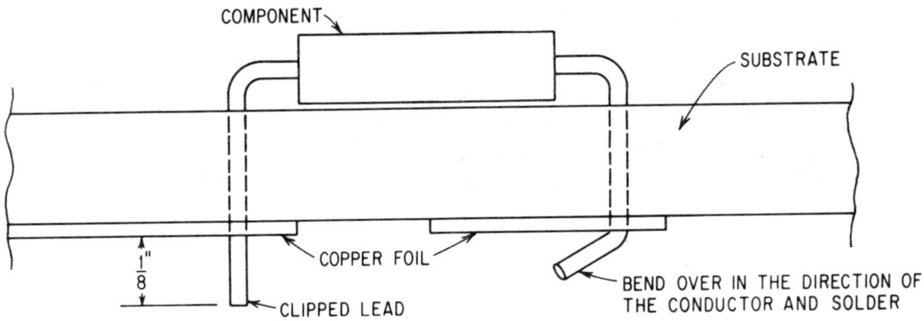

Figure 63 Leads clipped before soldering

left in the solder and that the pads are covered. Refer to Figures 62 and 63 and to the *Printed-Circuit-Board Soldering* assignment.

13. Complete any electrical tests on the board or perform other operations that would be appropriate.
14. Clean the board, put your name on it, and turn it in to your instructor.

CABLE-TYING SUBASSEMBLY (1)

In the wiring of chassis assemblies quite a number of special devices or special techniques are used. Some of these will be illustrated in this assignment.

The first of these is the use of the solderless connector. It has been said that this type of connection is better than a solder connection. It is quick and easy to make, is versatile, and when done correctly is excellent in terms of electrical conductivity and durability. Figure 64 shows a number of different types of solderless connectors.

The connection is made by stripping the insulation the correct distance, applying the connector to the wire, and then crimping the shank of the connector onto the wire using the proper tool. Figure

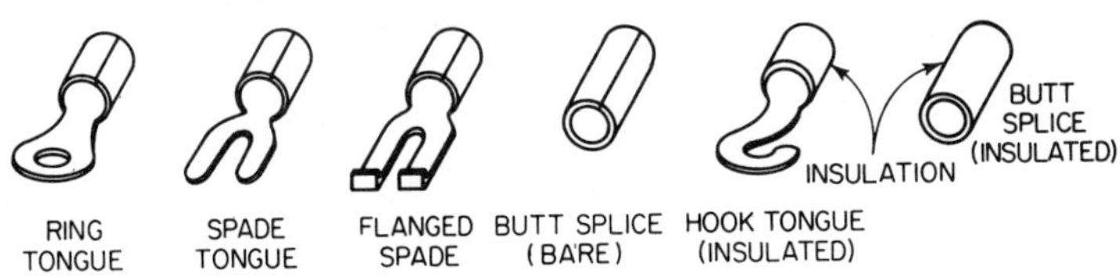

Figure 64 Solderless connectors

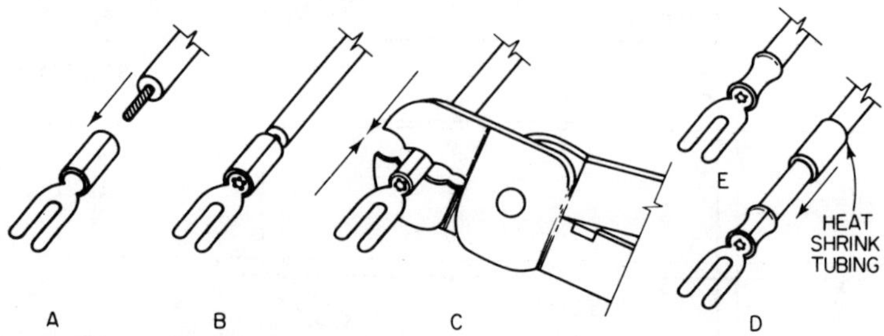

Figure 65 Making a solderless connection

65 shows this sequence. Often heat-shrink tubing is put over the shank for insulation and for improving the physical strength of the wire at that point. Many of these types of connectors also come with insulation on the shank, and the crimping tool squeezes through this insulation.

These connectors are sometimes electrically tested under operating conditions with a "pull" and a "voltage-drop" test. For a certain wire size, a specified current is fed to the connector. The voltage drop across the connector is measured before and after the connector is subjected to a specified pull stress. Refer to Appendix C for information on this type of test.

Splices usually are not made in electronics wiring. If a splice must be made, the wires would normally be run to a terminal point and both soldered to the terminal. In those rare cases when a splice is required, it should be done as illustrated in Figure 66.

The use of heat-shrink tubing was mentioned above in the examples of the solderless connector and the splice. This tubing is a special type of material that will shrink when heated. The tubing is formed in a state where the molecules of the material are extended out from each other, and when heated the molecules close back together, giving the effect of shrinking. The material is very tough and durable. As well as acting as insulation it also adds physical strength to a wire

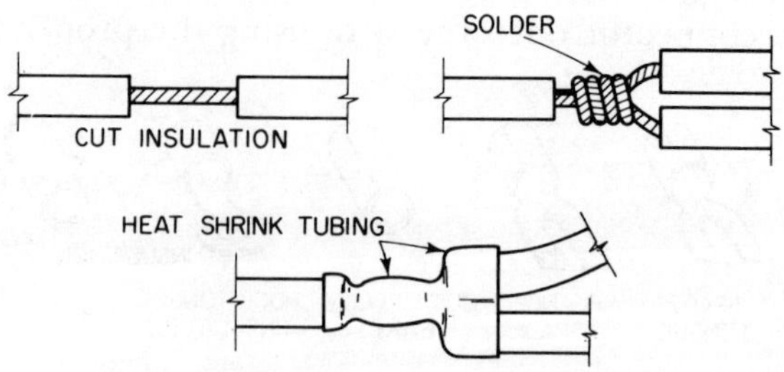

Figure 66 Making a splice

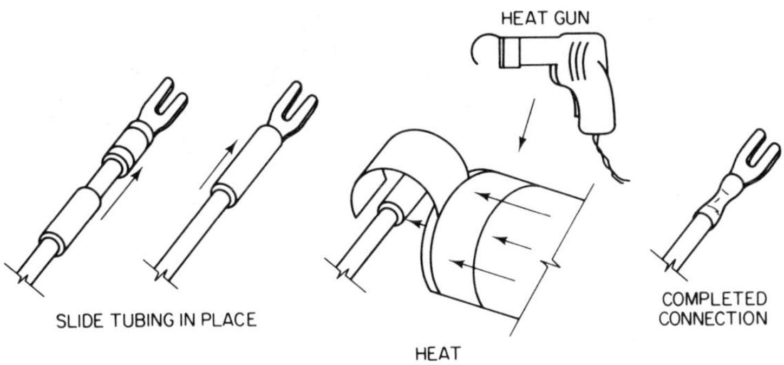

Figure 67 Application of heat-shrink tubing

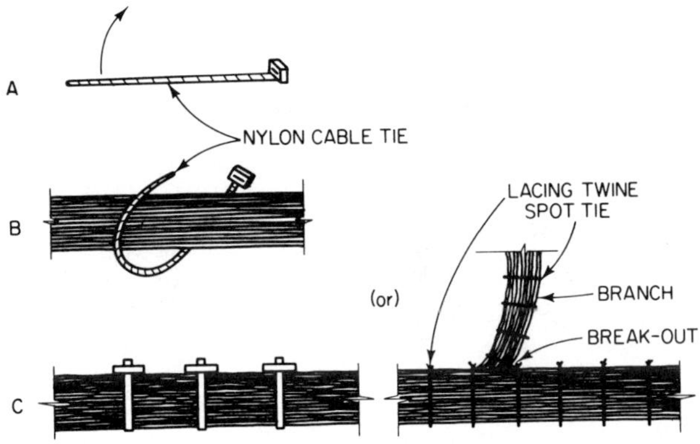

Figure 68 Harness spot ties

at the point of a connection. Figure 67 shows the application of heat shrink tubing over the shank of a solderless connector that has been squeezed on the wire.

Connections at various points along a cable run are made by certain conductors "breaking out" of the main bundle into a branch including one or more wires. The wires in the main cable run, and in the branches, are held tightly together with cable ties, spot ties, or in some cases with a lacing run. In most cases today, patented nylon cable ties are used or a special spot tie is made using a nylon lacing twine. Figure 68 shows the use of these tying arrangements. A tie should be on each side of a breakout.

Two types of spot ties are made using the nylon lacing twine. These are shown in Figures 69 and 70. One is made with a *clove hitch* and the other is made with a *loop,* referred to as a *girth hitch.* There are perhaps advantages or disadvantages for both types. With both ties, the lacing twine is cut and sealed by melting the nylon close to the knot with a soldering iron.

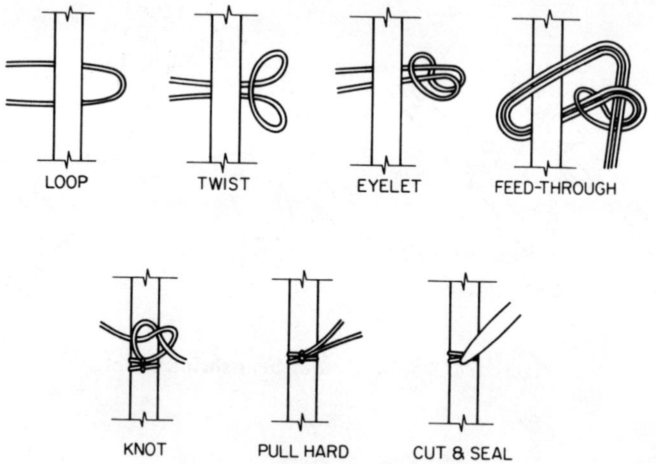

Figure 69 Girth hitch spot tie

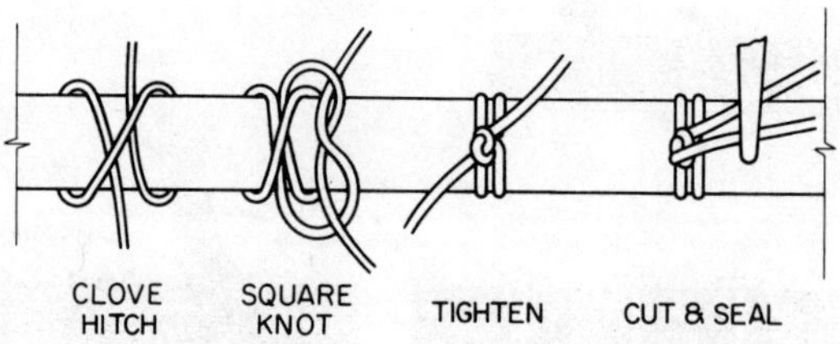

Figure 70 Clove hitch spot tie

Purpose of Assignment

To become acquainted with the following:

a. Use of a solderless connector.

b. Use of heat-shrink tubing.

c. Making a splice.

d. Spot ties.

e. Use of sleeving.

f. Wire color-code identification.

Tools and Equipment

Various hand tools for soldering.

Soldering device with dampened sponge.

Soldering vise.

Crimping tool.

Heat gun.

Goggles for eye protection.

Materials

Solderless connector.

Miscellaneous conductor wire—as required by instructor.

Heat-shrink tubing—2″.

Nylon lacing twine.

Sleeving—13″.

18-gauge bare wire—13″.

Liquid flux and isopropyl alcohol or flux remover.

20-gauge rosin-core solder.

Assignment

Make up a cable subassembly as indicated in the Directions and including the conductors specified by the instructor.

Directions

1. Make a straight cable subassembly, as shown in Figure 71. In the cable include approximately the following conductors, or those required by your instructor (refer to the *Wiring Harness Assemblies* assignment for information on the color code):

   a. 1—orange with yellow tracer, 24 gauge.

   b. 1—yellow with black tracer, 24 gauge.

   c. 1—purple, 18 gauge.

   d. 1—bare wire covered with sleeving.

   e. 2—white, 18 gauge (see step 5 in Directions).

   f. 1—black, 22 gauge (see step 5 in Directions).

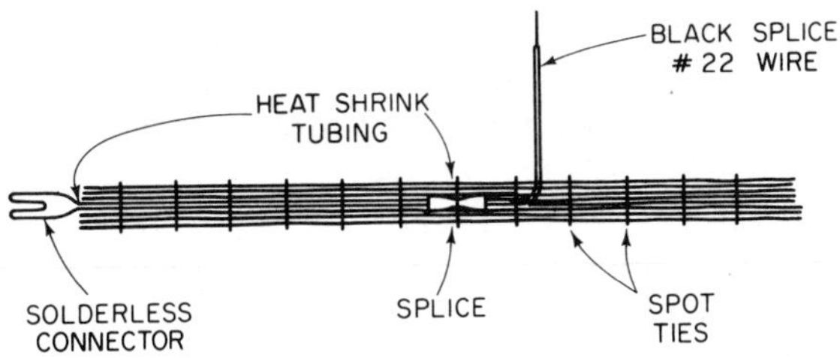

Figure 71 Cable subassembly

2. Cut the conductors to exactly 12″.
3. Strip all conductor ends to exactly 1/2″ and tin the conductors using proper soldering techniques. Use a heat sink.
4. Use the sleeving over the bare conductor.
5. Prepare two 12″ 18-gauge white wires as follows:
   a. Crimp the solderless connector on one end of one of the 18-gauge white wires. Refer to the introduction.
   b. Splice a 5″ black 22-gauge wire on the other 18-gauge wire. Make the splice 4″ from the end of this wire. Refer to the introduction.
6. Cut the heat-shrink tubing into 3/4″ and 1¼″ pieces. Shrink the short piece over the shank of the solderless connector and the other piece over the splice made in step 5b.
7. Assemble the wires together into a straight cable except for the black wire splice as shown in Figure 71. Hold the wires together with temporary ties until the permanent ties are put in place. This will also allow the lay of the wires to be adjusted so that all wires run parallel and straight.
8. Spot-tie the cable every 3/4″. Use nylon lacing twine and tie the cable up to 1½″ from the ends of the cable. Refer to the introduction for information on the two types of spot ties. Tie half of the cable by the clove hitch method and the other half of the cable by the girth hitch method.
9. When the cable is finished, put your name on it and turn it in to your instructor.

WIRING HARNESS ASSEMBLIES (1)

When numerous wire connections have to be made in a chassis, the wires will often run parallel to each other for at least part of the distance. Definite wire routes can easily be established, and a perfect situation develops for the making of a pretied harness, or cable, to fit the wiring requirements. Figure 72 shows an example of a typical pretied cable.

This type of cable harness assembly can be made on a production-line basis. Not only is time saved in the actual assembly of the harness but also in correcting wiring errors that would otherwise occur. Space is usually conserved, and a neat and uniform wiring layout is also assured.

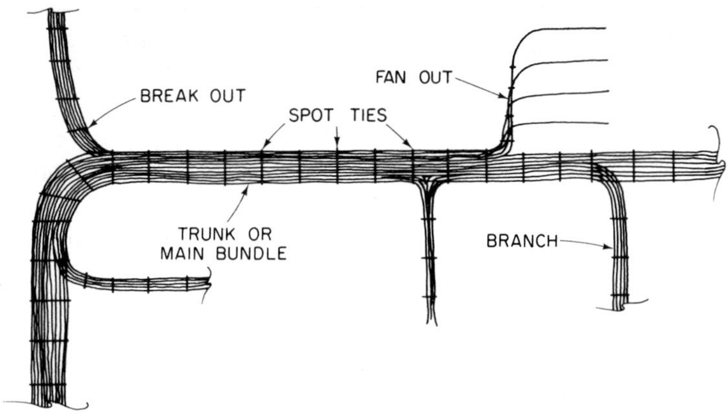

Figure 72 Typical wiring harness

For making the cable, precut wires are assembled according to a predetermined plan. The wires are formed together on a harness board that has been designed so that the cable is routed to fit a particular circuit assembly. Connections at various points along the cable run are made by certain conductors "breaking out" of the main bundle into a branch, including one or more wires. The wires in the main cable run, and in the branches, are held tightly together with cable ties, spot ties, or in some cases with a lacing run. In most cases today, patented nylon cable ties are used or a special spot tie is made using a nylon lacing twine. Refer to the introduction of the *Cable Tying Subassembly* assignment for information on the use of these tying arrangements. A tie should be on each side of a breakout.

The spot ties should be spaced at a distance equal to twice the cable diameter or 1/2″ minimum spacing. A maximum spacing of three times cable diameter is recommended. There should be equal spacing between ties and also a tie should be on each side of a breakout. While tying, the lay of the wires in the bundle should be kept straight and even. Crossovers should be avoided except where the wires exit from the bundle.

The wires for the cable are prepared ahead of time according to specifications, including size, type, length, ends stripped, ends tinned, lugs mounted, and markings or identification labels attached. A wire table or wiring sequence list is used to show this information.

Insulated wire used for harness or electronic equipment wiring is very often color-coded. Following the normal color code for the numbers 0 to 9, four possible code arrangements may be used, as follows:

Number of Digits in Code		*Wire Colors*
a.	1	Solid color insulation
b.	2	First digit is background color Second digit is color of stripe
c.	3	First digit is background color Second digit is wide stripe Third digit is narrow stripe
d.	4	First digit is background color Second digit is wide stripe Third digit is first narrow stripe Fourth digit is second narrow stripe

As an example, if the code is 92, the background of the wire is white and there is one red stripe. If the code is 420, the wire is yellow with a wide red stripe followed by a narrow black stripe. The color stripe system is illustrated in Figure 73.

Using the three-digit arrangement, 810 possibilities exist; using the four-digit arrangement, there are approximately 10,000 possibilities.

To facilitate the making of the harness, a wiring harness board is made up to match the requirements of a particular chassis. The board is usually wood, with guide pins which are specially designed harness board nail. These guide pins are located at bends in the cable run, corners, and breakout points. The ends of the wires may be secured by pins, clips, or springs mounted on the board. A typical harness assembly drawing is shown in Figure 74. In some cases 90° bends in the vertical direction can be made by using a hinged surface on the board.

Purpose of Assignment

1. To become acquainted with color codes.
2. To learn how to make cable spot ties.
3. To learn how to make a wiring harness.

Tools and Equipment

Various hand tools for soldering.

Soldering device with dampened sponge.

Soldering vise.

Harness wiring board with harness assembly drawing (Figure 75).

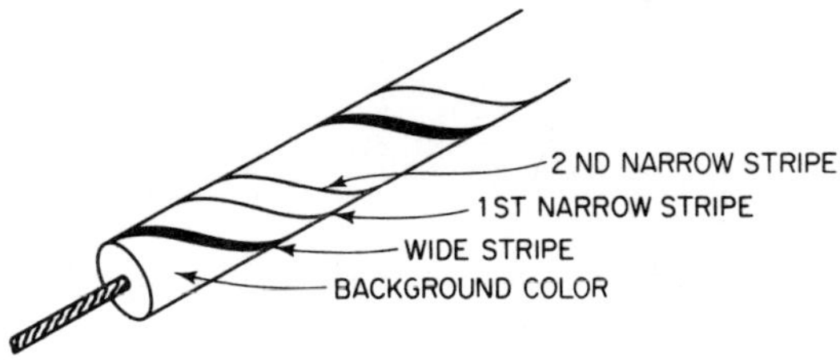

Figure 73 Wire color code

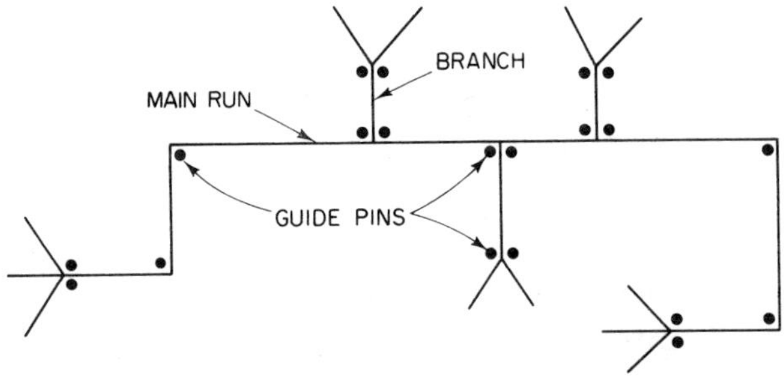

Figure 74 Harness assembly drawing

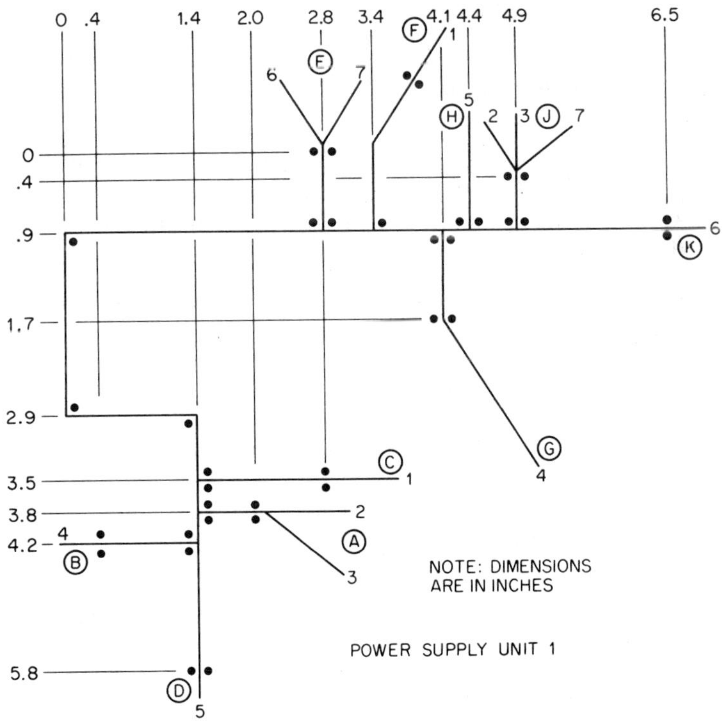

Figure 75 Harness assembly drawing

Materials

Nylon lacing twine.

Conductor wire as listed in the wiring sequence list (Figure 77).

Masking tape 1″ wide—6″ long.

Liquid flux and isopropyl alcohol or flux remover.

22-gauge rosin-core solder.

Assignment

Make up a wiring harness assembly for the power supply, Unit 1 (Figure 75). (See the Directions.)

Directions

1. Read the information in the introduction.
2. Refer to the information in the wiring sequence list, Figure 77. Precut the required wires.
3. Strip the insulation exactly 1/2″ and tin each end of the conductors. Use a heat sink.
4. Label each conductor with its wire identification number. Put the label on both ends of each conductor. Refer to Figure 76 for the labels.
5. Install the wires on the harness board. Follow the wiring sequence list. Wires should run parallel to each other in the bundle and should not weave in and out except that branched leads may cross other conductors.
6. Use spot ties to tie the bundle. Use either type of tie explained

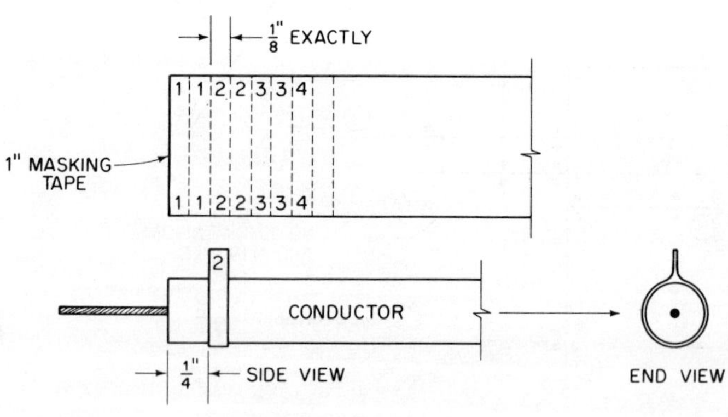

Figure 76 Conductor labels

WIRING SEQUENCE LIST
Power Supply Unit 1

ITEM NO.	ROUTE	WIRE IDENTIFICATION NO.	AWG	COLOR	LENGTH (INCHES)
1	C–F	1	18	999	12
2	A–J	2	18	777	12 $\frac{1}{4}$
3	A–J	3	22	990	12 $\frac{1}{4}$
4	B–G	4	22	915	13
5	D–H	5	22	000	12 $\frac{1}{4}$
6	E–K	6	18	222	6
7	E–J	7	22	990	5

Figure 77 Wiring sequence list

in the introduction. Pull each of the ties tight when making the tie.

7. Keep the cable formed exactly to the shape of the layout. Make the corners sharp and square. Use temporary ties as needed.
8. Lift the cable off from the board, reshape it and retie any loose ties.
9. Put your name on the cable when it is completed and turn it in to your instructor.

PART TWO

PRINTED-CIRCUIT ARTWORK

3

Techniques and Development of Printed-Circuit Artwork

SPIRAL-SWEEP-GENERATOR ARTWORK (1)

As an introduction to printed-circuit artwork, this assignment incorporates a few of the fundamentals of what is involved in this phase of circuit fabrication. The basic concepts of the development of the pattern and the use of a light-sensitive film are introduced. More detailed instructions and various techniques will be given in other assignments.

The pattern for the printed circuit is made up from the schematic drawing of the circuit. The arrangement of the parts in the circuit needs to be studied and the best location for those parts on the printed-circuit board needs to be determined. In positioning the parts, the factors to be considered are the size, weight, physical structure, electrical relationships, heat dissipation, and conductor pattern conflicts. An evaluation of these factors should be made in terms of their relative importance in each individual situation and a decision should then be reached on the final location of the parts. Usually several trial layouts are made on paper. Parts and conductors are shifted and moved to improve or simplify the circuit pattern. After working out any flaws in the pattern, the layout is then drawn to an exact scale and the location of each item is finalized.

In making the preliminary layout, all junction points should be labeled for identification to make it easier to trace back and verify its accuracy as related to the schematic drawing. The layout should be planned to include a pad or land (see the Glossary in Appendix A) for each connection to the copper (printed-circuit pattern) trace (refer to Figure 39).

It must be noted that a pattern can be viewed from the front or the back side of the board. Depending on how the layout is viewed, the final board could inadvertently be made to face backward or upside down. Initially, it is probably best to visualize the layout the way the layout would appear as viewed from the *component* side of the board. Following this system the input to a circuit will end up on the left and the output on the right, the board will be made to fit correctly in an edge connector, or the board will simply fit into the circuit where it is to be installed. From the *conductor* side of the board the layout will be a *reverse* pattern, and when the actual board is exposed this reverse pattern must be used.

The spiral-sweep-generator circuit is a "mixing" circuit for two signal frequencies and is used in conjunction with an oscilloscope. Depending upon the frequencies, $F1$ and $F2$, various amplitude and phase combinations will appear at the vertical and horizontal outputs. When connected to an oscilloscope the signals will determine vertical and horizontal deflection, resulting in very many intriguing and interesting sweep patterns. For a person with an interest in or a knowledge of oscilloscopes, this is a worthwhile device. Many hours of time can be spent studying frequency relationships (refer to Figure 78).

In this assignment a printed-circuit photomaster will be made, and a film negative will be printed from the photomaster. To make the photomaster, layout tape is applied onto a sheet of plastic or vellum which overlays the proposed circuit pattern. This photomaster can be made at an expanded scale or, as in this assignment, it can be made at a 1:1 scale. A negative is then made from the photomaster. This negative, in conjunction with a photosensitive emulsion, is used to obtain a resist pattern on a printed-circuit board (refer to Figures 86 and 87).

Purpose of Assignment

1. To gain experience in developing a printed-circuit layout.
2. To gain experience in printed-circuit photographic techniques.

Tools and Equipment

Various hand tools.

Pocket knife, razor blade, or X-acto knife.

Exposure printing frame with proper ultraviolet light source.

Materials

Masking tape.

Grid paper (0.1″ squares), scratch paper, pencil, and eraser.

Sheet of vellum, clear plastic, or tracing paper.

1′ of 1/16″ printed-circuit layout tape.

5—1/4″ printed-circuit layout pads.

12—3/16″ printed-circuit layout pads.

3M Color Key film and developer.

Assortment of press-on letters.

Assignment

Make a printed-circuit photomaster and a negative of a circuit to be constructed on a printed-circuit board.

Directions

1. Obtain the circuit drawing for a circuit to be constructed. This can be any simple circuit or the circuit for the generation of a spiral sweep on an oscilloscope such as the one shown in Figure 78.
2. Make a trial layout of the circuit with pencil and eraser.
   a. Determine the basic arrangement of the parts of the circuit. In this case use one line of components going across the board from left to right. Orient the components parallel to each other and parallel to the left and right edges of the board. Observe clearance and spacing specifications as required.
   b. Consider whether the layout is viewed from the component side or the conductor side of the board. In this case make the pattern layout from the *component* side of the board.
   c. Label all *electric* junctions on the circuit drawing with a letter or number.
   d. Make a rough sketch of the conductor pattern. Use a pad for each connection to the pattern. Draw in the symbols for the components and label the junctions with the letter or number from step c.

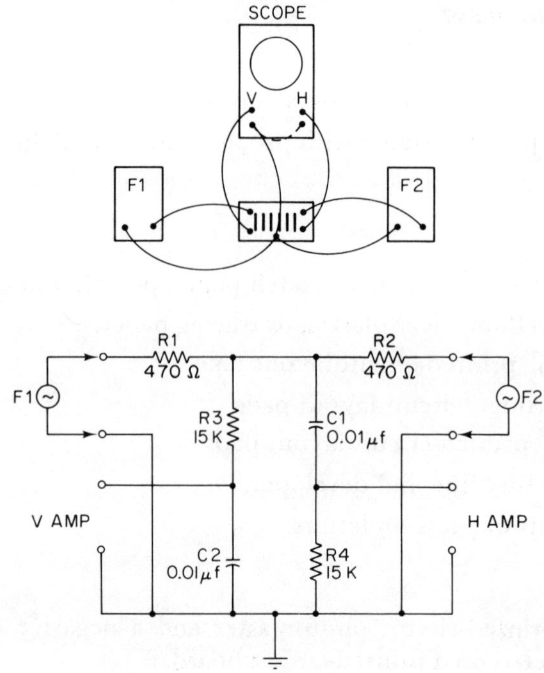

Figure 78 Spiral sweep generator circuit and connection diagram

e. Evaluate the layout in terms of the complexity of the conductor pattern and the location of the components.

f. Determine if there is a way of simplifying the layout. Eliminate all crossovers.

g. Redraw the layout as many times as required to make a sensible and workable arrangement of the components. Use a pencil and an eraser.

3. Redraw the layout again to an exact 2:1 scale. Use the layout arrangement and specifications given in Figure 79. Double the exact size of each component. For the purpose of this assignment consider the resistors to be 3/16 X 9/16″ and the capacitors to be 1/4 X 5/8″. This layout is made only for the experience of doing an expanded scale layout.

   a. Locate the terminals according to the dimensions and layout arrangement given in Figure 79.

   b. Locate the pads 0.2″ from the end of the body of the components (0.4″ at a 2:1 scale).

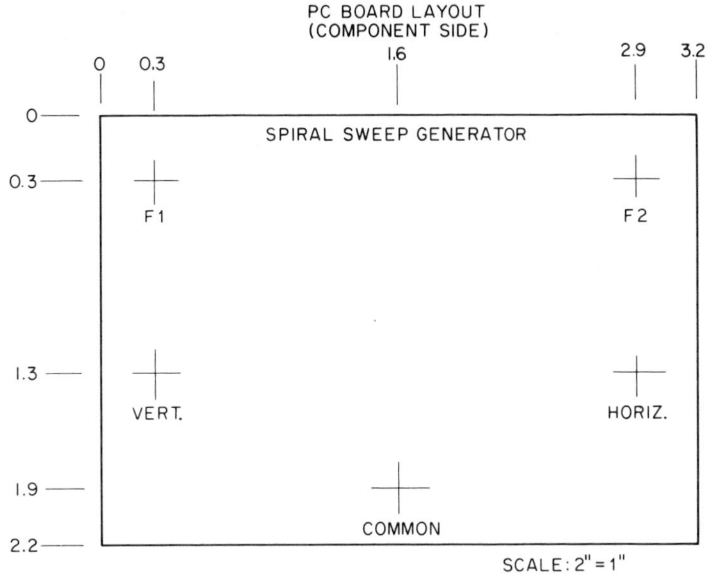

Figure 79 Printed-circuit board layout specifications

4. Using the exact size of each component make an exact layout at a 1:1 scale. Use a pencil and eraser.
   a. Use a grid paper to keep the dimensions accurate, the pattern lined up straight, and the components accurately placed (0.1″ squares).
   b. View the pattern from the component side of the board.
5. Make a photomaster on a piece of vellum or clear plastic.
   a. Tape the vellum or plastic over the circuit layout on the grid pattern made above (1:1 layout).
   b. Place the pads in place, lined up with the appropriate grid lines. Use an X-acto knife to hold and carry the pads to the place where they go on the pattern (refer to Figures 86 and 87).
   c. Following the pencil layout of the circuit pattern, put the conductors in place between the pads, using the printed-circuit layout tape. Guide the tape with the X-acto knife and cut the tape by holding it down with the knife edge and jerking the tape.
   d. Mark the cut lines of the board at the corners using short strips of printed-circuit layout tape.

e. Using press-on letters, put your initials on the photomaster. In doing this, view the photomaster as it would be seen from the *conductor* side of the board. This is a reverse from that used above.

6. Using the photomaster made above make a 3M Color Key film negative. Expose from the conductor side of the photomaster.

   a. Follow the directions given on the film package or refer to notes on a demonstration of how to use the film. Develop, clean, and dry the film carefully. In these operations be very careful to rub the emulsion very lightly. Do not scrub it hard but wipe it gently.

7. Turn in the 1:1 pencil layout, the taped photomaster, and the Color Key film negative.

PRINTED-CIRCUIT ARTWORK DEVELOPMENT

Printed circuits are used extensively today and will be used for many years to come. This type of construction does not solve all circuit construction problems but does have the advantages of reproduction accuracy, compactness, and adaptability.

The printed-circuit board is a fabricated laminate with a thin copper foil bonded to it on one or both sides. This copper foil is processed in certain ways to form a pattern that follows the requirements of a circuit. The components are mounted on the front side of the board, and the leads are fed through the board and soldered to the conductor pattern on the back side. The design of the conductor pattern and the development of the necessary artwork is the subject of this assignment. It should be pointed out that there are many variations and individual preferences in the development of printed-circuit artwork and that the author's presentation is just one of several systems that could be used.

There are many items relating to the artwork detail, but the greatest problem is the arrangement of the components and the development of the conductor pattern within a specified board area. This problem can be solved in a number of ways. Basically, it can be solved by calculating in some way the board area required (depending upon the number and size of components) and then fitting the components into that area. It can also be solved by cutting out "paper-doll" templates (shapes of the components scaled to size), fitting them together, and then determining the required board area.

Both basic systems perhaps have merit, but the first system seems best suited to the technician.

Generally, in the construction of printed-circuit boards, after a preliminary examination of the construction possibilities of the circuit, it must be decided how the board is to be mounted and how electrical connections to the board are to be made. Usually, the external wires coming in to the board are connected to terminal posts on the board, make connection through an edge connector that fits the board, or, in rather poor quality construction, are soldered directly to the conductor pattern. This assignment will involve the use of the edge connector, which is illustrated in Figure 80.

The overall artwork incorporates a great deal of detail, which is brought out in varous drawings, sketches, or taped-up arrangements of the conductor pattern. There is a layout sketch showing the components and the conductor pattern or the components alone. There are drawings that show the placement of holes, board dimensions, and the labeling of components for identification. Most important is the photomaster, which is made up of sharply defined pattern lines (printed-circuit tape) and markings, and usually contains no other detail except what might be directly related to the copper pattern. The photomaster is photographed and the resulting film negative is used in conjunction with a light-sensitive emulsion in order to get a resist pattern on the board prior to etching the board.

As a starting point in the development of drawings and artwork for a printed circuit, it is necessary to determine the approximate minimum board area required for the particular circuit. This is done by finding the area required for each component, adding up the total area, and then adding 10% for spacing requirements.

From Figure 81 it can be seen that the distance between mounting-hole centers is equal to the sum of the body length, lead diameter, and 0.08″ on each side of the component. This allowance is made for the purpose of meeting machine-handling requirements on the

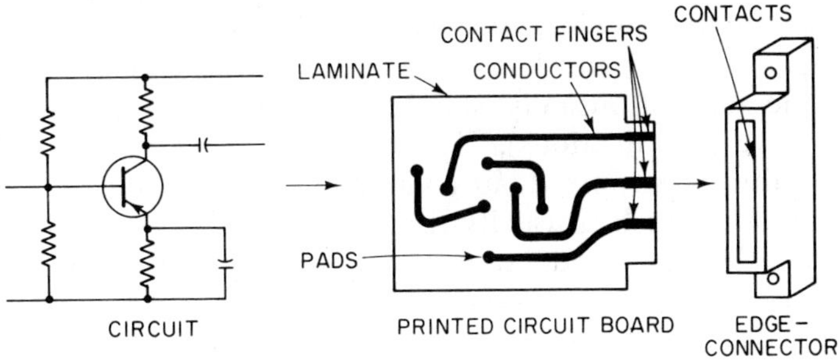

Figure 80 Printed-circuit board edge-connector

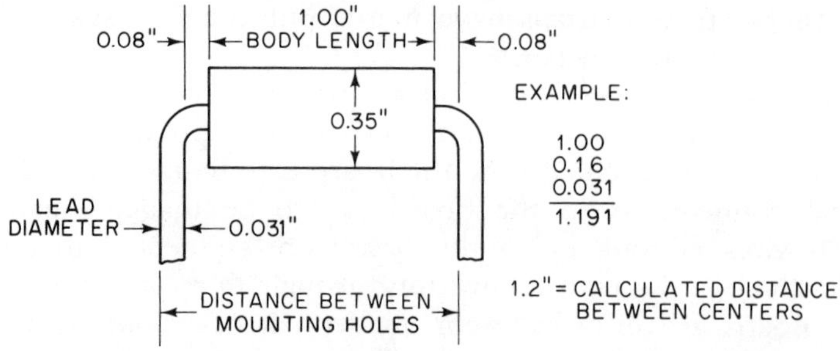

Figure 81 Calculation of distance between mounting holes

production line. The figure arrived at from this addition is rounded off to the nearest tenth of an inch. The component can then be drawn to size on a grid system (0.1" between lines) and its area in square tenths of an inch determined. For adequate allowance, at least a full tenth of an inch of space must surround all parts of the component. The rectangle representing the area accountable to that component is thus defined. This is shown in Figure 82. The area for sockets and other components is determined in a similar manner. For components the same size, find the area for one and simply multiply by the number of components to get the total area of those components, add the area for the other components, and then add 10% to get the *grand total approximate minimum* board area for a particular circuit.

After obtaining the area, the length and width of the board can be determined. For example, if the board has 6 square inches of area and it must fit into an edge connector strip 2" wide, it must be 3" long.

Machine indexing holes are located in a board to facilitate automatic assembly. The centers of these holes are oriented with the grid lines upon which the board layout is sketched. Usually, the holes are located along the longer dimension unless the length and width are 6" or less and the minimum spacing of the index holes is approximately two-thirds of the longer dimension. The diameter and location of the holes is usually standardized (refer to Figure 83 for representative specifications). Note the area surrounding each hole, where circuitry is permitted but components are prohibited.

The layout sketch is usually drawn double size or to some other expanded scale and a reduction to normal full size is made in the photographic process of making the film. This is done to increase detail, line definition, and accuracy. In making the sketch, the components, conductors, and pads are also expanded to the same scale.

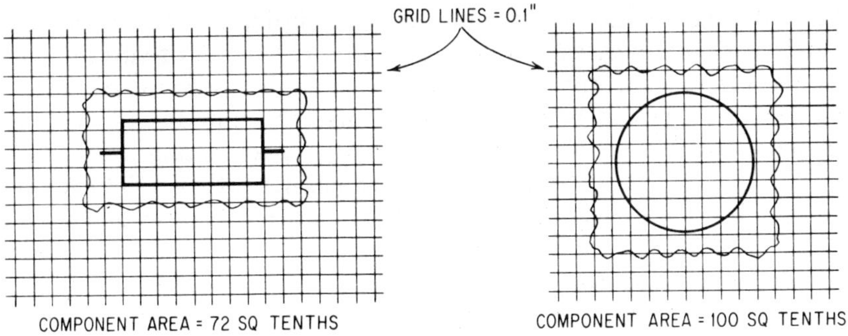

Figure 82 Component-area calculation

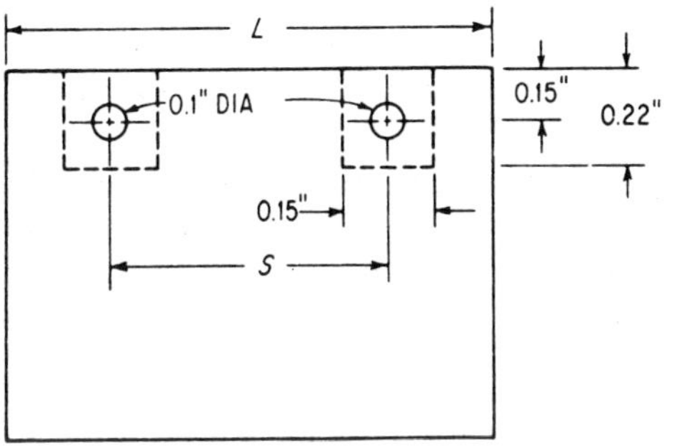

L DIMENSION		S DIMENSION
FROM	TO	
1.2	1.9	0.8
2.0	2.8	1.5
2.9	3.5	2.5
3.6	4.4	3.1
4.5	5.3	4.1
5.4	6.2	5.0
6.3	7.0	5.9
7.1	8.5	6.7

Figure 83 Dimensions for index holes

A board *identification number* is usually etched in the copper off to one corner of the board. This number should show on the drawings. *Registration marks* or datum points (usually three) will also appear on all drawings and are used for the purpose of aligning the drawings when they overlay one upon the other. When a drawing is to be photographed and reduced in size, *photoreduction targets* should be used. These are simply marks set up to be representative of a certain dimension after the reduction is made. The numbers used for this distance are usually whole numbers (refer to Figures 87, 88, and 89 for examples of these items).

In sketching the proposed layout pattern avoid crossovers whenever possible. Usually with a little study and reworking, most crossovers and most pattern complications can be eliminated. If it is necessary, a crossover is made by going under the place where a component is mounted, by going around the end, or by using a jumper wire fed through from the component side of the board.

The width of the conductors and the spacing between conductors is somewhat arbitrary because it is usually very easy to have more than adequate width or spacing. For example, a 1/16″ conductor, if it is 1-oz. copper, can carry approximately 0.5 A of current. Most printed circuits do not have anywhere near that amount of current. As for the spacing between conductors, the amount of electrical capacity will be about 1 pf/in. for 1/32″ spacing, and a spacing of 1/16″ assures a voltage breakdown capability of over 500 V (ac). Except for high-frequency or feedback problems, the question of width or spacing centers around the practical matter of fabrication. It is more difficult to make the conductors very narrow or very close together. It is recommended, then, that a conductor 1/16″ wide and a minimum spacing of 1/32″ be used for general circuit work.

The spacing to mechanical portions of the board is important. At least 0.1″ clearance should be kept between the conductors and screw heads, mounting brackets, and other items. Drafting mediums should be dimensionally stable in order to hold to accurate requirements.

Perhaps many new terms have been used in this introduction and many of these terms will be encountered as a person becomes more involved in this particular area of work. A glossary of terms and definitions is given in Appendix A. Also refer to the *Layout Development Check List* in Appendix A.

Purpose of Assignment

1. To gain experience in the development of printed-circuit artwork.
2. To gain experience in making a printed-circuit photomaster.
3. To gain experience in making the drawings used in circuit-board construction.

Tools and Equipment

Measuring scale.

Tweezers, razor blade, or X-acto knife.

Micrometer.

Materials

Drawing paper, pencil, and eraser.

Grid paper (0.1″ squares).

Printed-circuit tape and pads.

Sheet of vellum, clear plastic, or certain kinds of tracing paper.

Masking tape.

Assignment

The circuit of Figure 84 is to be built onto a printed-circuit board which will fit into an edge connector.
Prepare the following drawings:

1. A photomaster of the conductor pattern.
2. A drawing of the component placement.
3. A hole-layout and board-dimension drawing.
4. A component-labeling drawing.

Directions

The following is a general outline for the development of printed-circuit artwork. Some of the specifications and suggestions in this outline are somewhat arbitrary and could be decided upon on the basis of individual preference—such as, for example, the question of whether the conductor corners should be rounded or square, or the decision as to the width of the conductors.

1. Make a preliminary evaluation of the layout as to the approximate location of the parts and the arrangement of the connections to the edge connector. Evaluate each component as to its size, weight, heat dissipation, mounting facilities, and whether or not it is controllable. Make an attempt to foresee any conflicts in the conductor pattern in terms of crossovers, spacing, or unnecessary complications.
2. Calculate the *approximate minimum board* area required for the given circuit.

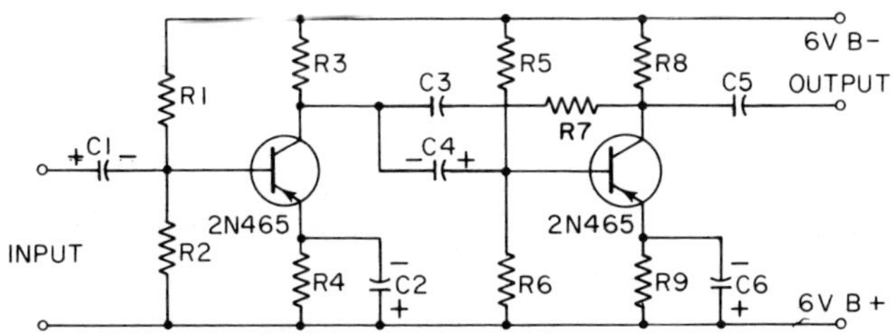

ALL RESISTORS ARE 0.5 WATT
C_1, C_2 25 VOLTS, ELECTROLYTIC (5/16" DIA. x 3/4" LONG)
C_3, C_4 6 VOLTS, ELECTROLYTIC (1/4" DIA. x 5/8" LONG)
C_6, C_5 200 VOLTS, PAPER (3/8" DIA. x 1" LONG)
MOUNT IN EDGE-CONNECTOR WITH CONTACT FINGERS
USE A TRANSISTOR SOCKET FOR TRANSISTORS

Figure 84 Fixed-gain amplifier

a. Measure the size of each component. Record the information. Measure the pigtail-lead diameters with a micrometer. Use an ordering catalog to obtain this information if the parts are not available.

b. Determine the distance between the mounting-hole centers for the components mounted by pigtail leads. Round off to the *nearest tenth* the sum of the body length, a machine-handling allowance of 0.16″, and the lead diameter (refer to Figure 81). Record and repeat for each lead-mounted component.

c. Calculate the area required for each component (refer to the introduction for information).

d. Make multiplications and additions to determine the total area required by the components. Add 10% to the area to allow for spacing and convert to square inches by dividing by 100. (This gives the approximate minimum board area in square inches.)

3. Determine the length and width of the board to be used. The dimensions for the edge connector are shown in Figure 85.

4. Set up the contact fingers of the pattern to make the following connections in the edge connector, as viewed from the conductor side of the board, and reading from *right to left:*

Pin 1— common
Pin 3— input
Pin 10— output
Pin 12— B—

5. Determine the location of the machine-indexing holes if they are to be included (refer to the introduction and Figure 83 for information relating to the dimension requirements).

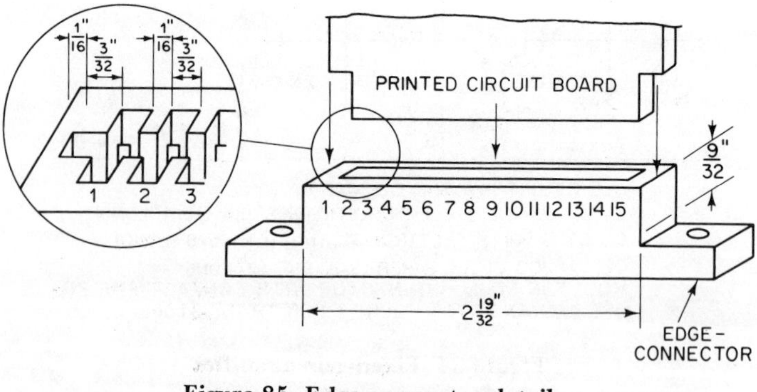

Figure 85 Edge-connector detail

6. Using a soft pencil that can be erased, make up a layout drawing of the circuit pattern on paper. Show the pattern of conductors as viewed from the conductor side of the board, the placement of the components, contact fingers, pads, and the board size and shape all drawn to an exact *double scale.*

   a. First make a *trial layout* of the pattern. Label junctions for identification. Run conductors as short and direct as possible and round the corners. Avoid crossovers and conflicts in the conductor pattern.

   b. Use a pad or land for *each* connection to the board and use sockets for mounting transistors.

   c Keep similar parts oriented to face in the same direction, or at right angles to the edge of the board. Consider the *electrical* and *physical* relationship of parts.

   d. Allow for at least 1/16″ clearance between components, and between the edge of the board and conductors. This reduces to 1/32″. Keep at least 0.2″ clearance to mechanical parts such as screw heads or mounting brackets. Avoid large areas of conductor pattern, unless the areas are relieved to prevent blistering.

   e. *Rework, rework,* and *rework* the layout into its simplest form. Start over completely if necessary to obtain the simplest and most direct pattern.

7. From the layout drawing just completed make up a double-sized photomaster as viewed from the printed-circuit side of the board. Use a sheet of vellum or clear plastic as an overlay on top of the layout sketch as shown in Figure 86. Place 1/4″ round *pads* in place using tweezers and then run the *conductors* where they should be using a 1/8″-wide tape. On the photomaster also include *photoreduction targets,* short strips of tape marking off the *board outline,* and the board-identification numbers (see Figure 87).

NOTE: If the pattern is to be etched on printed-circuit board, a photographic film negative (reduced 50%) would be made from the photomaster. If photographic equipment is not available, the film can be made by a local commercial firm. For the purpose of this assignment the film need not be made. If the layout is done at a 1:1 scale, 3M Color Key film could be used as a means of obtaining a negative for circuit-board etching.

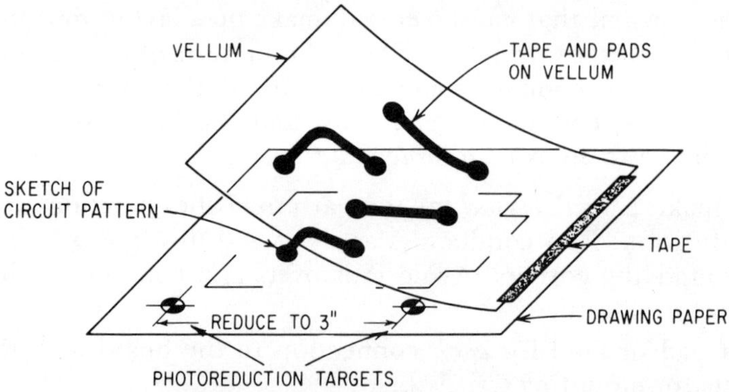

Figure 86 Double-sized photomaster layout

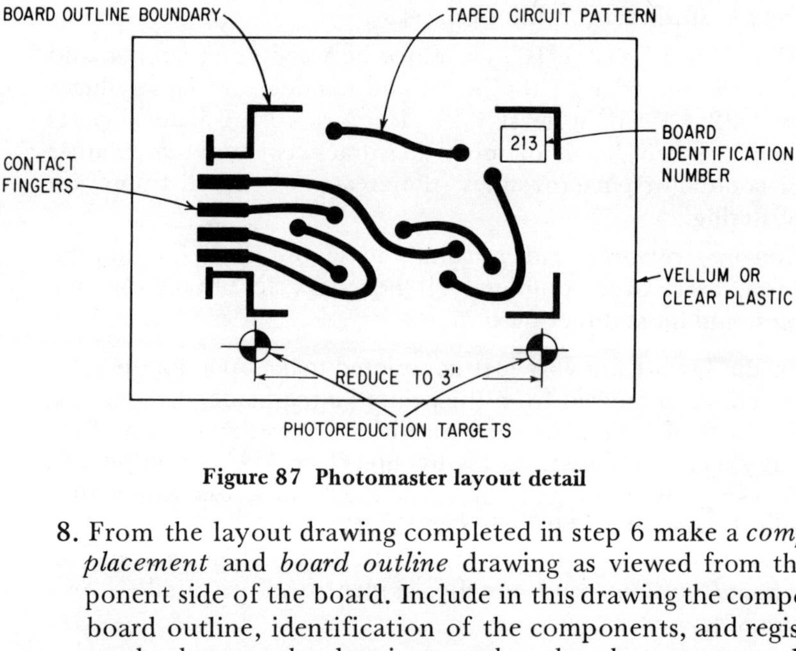

Figure 87 Photomaster layout detail

8. From the layout drawing completed in step 6 make a *component placement* and *board outline* drawing as viewed from the component side of the board. Include in this drawing the components, board outline, identification of the components, and registration marks; be sure the drawing matches the photomaster and layout drawing (see Figure 88).
9. From the sketch completed in step 6 make a drawing showing the *hole layout* and *board dimensions*. (See Figure 89 for an example.)
10. From the layout sketch completed in step 6, make a *component-labeling* drawing as viewed from the component side of the board. Include part numbers and symbols for diodes off to the edge of where the components will be located. Set up the part

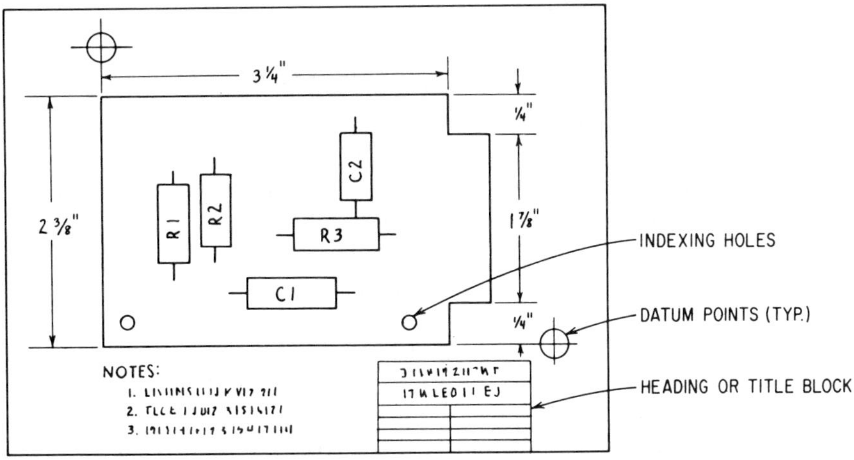

Figure 88 Component placement and board outline

numbers to read from *left to right* and from *bottom to top* (see Figure 89).

11. Include all necessary heading or title-block information on the drawings. Be sure that the board identification number (in this case a made-up number) appears within the board area, drawing registration marks appear in the proper place, photographic reduction targets are properly located, and that necessary dimensions are given.
12. When completed, turn in the photomaster and the drawings to your instructor.

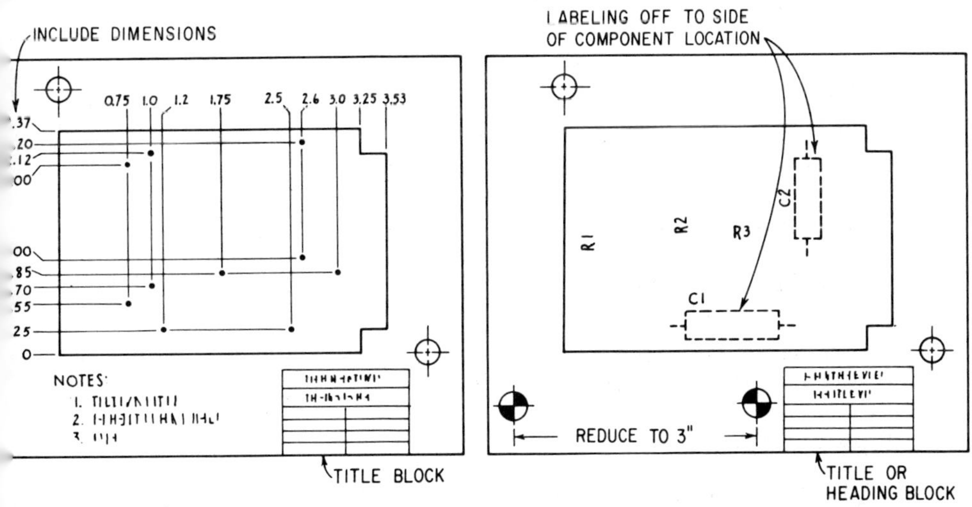

Figure 89 Hole-layout, board-dimension, and component-labeling drawing

As the components become more bulky or physically awkward, it becomes necessary to mount them off the board and on a related chassis. This is true with an item such as the battery, which, in addition to its weight, also needs to be specially mounted because of replacement problems. A push-button switch at times must be located on a panel where it is accessible. With these types of components, connections into the circuit must be made by running leads to the board from the place where the part is located.

In addition to the types of parts mentioned above, there are parts that require a special mounting arrangement for holding them to the board. An example is the type that incorporates the use of a tab of metal that slips through a rectangular slot to a special pad on the conductor side of the board. This tab can be bent over and soldered to the pad, or in some cases just simply soldered to the pad. Quite a number of components make use of this arrangement. In the case of the small transformer, it is held to the board by use of the tab, and the leads feed through the board to the appropriate part of the circuit. These leads are soldered to a pad at each connection point.

In the design of the layout of a circuit, the measurements, physical dimensions, and requirements of the components must be determined. This information is obtained by direct measurement, visual observation, the use of specification sheets, or by the use of parts-ordering catalogs.

The circuit in this assignment involves the use of components that are mounted in various ways. Their placement in a layout is determined by considering the factors mentioned in this and other assignments, and also by the use of common sense and good judgment.

Purpose of Assignment

1. To gain experience in the mounting of various types of components to a circuit board.
2. To gain experience in planning the layout and developing the artwork for the construction of a circuit on printed-circuit board.

Tools and Equipment

Tools and Equipment as listed in *Printed-Circuit Artwork Development.*

Materials

Materials as listed in *Printed-Circuit Artwork Development.*

Assignment

The circuit of Figure 90 is to be built onto a printed-circuit board which will fit into an edge connector strip. Prepare the following drawings:

1. A photomaster of the conductor pattern.
2. A drawing of the component placement.
3. A hole-layout and board-dimension drawing.
4. A component-labeling drawing.

Directions

1. Read the introduction to *Printed-Circuit Artwork Development.*
2. Follow the instructions given in the Directions of *Printed-Circuit*

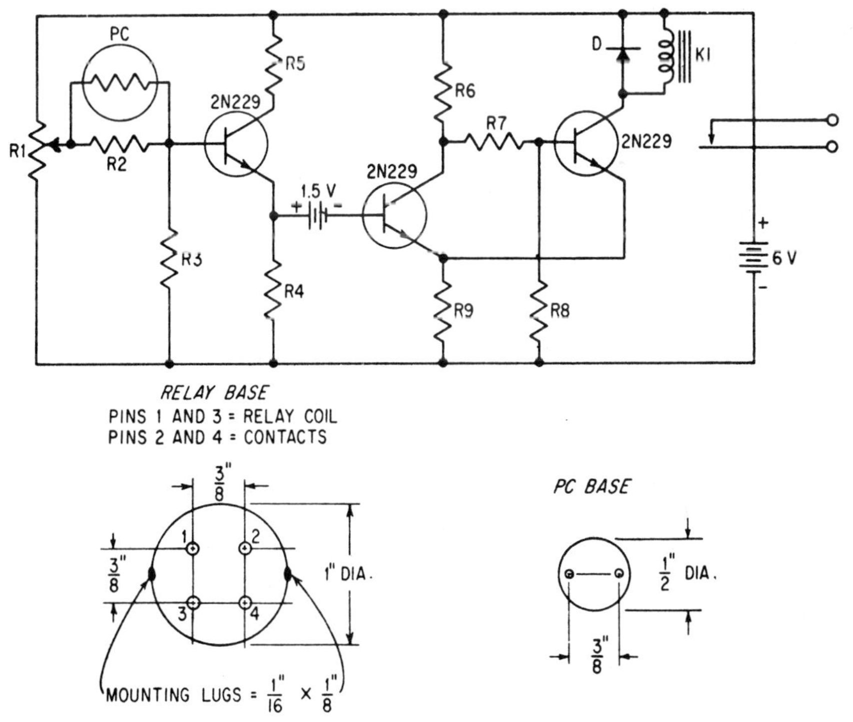

Figure 90 Light-operated relay circuit

Artwork Development except for step 4, and the *Layout Development Check List* in Appendix A.

3. Set up the contact fingers of the pattern to make the following connections in the edge connector strip as viewed from the conductor side of the board and reading from *right to left:*

   Pin 1— relay contacts
   Pin 3— relay contacts
   Pin 5— +6 V
   Pin 7— -1.5 V
   Pin 10— +1.5 V
   Pin 12— -6 V

4. When completed, turn in the photomaster and drawings to your instructor.

PART THREE

BASIC CHASSIS CONSTRUCTION

4

Preliminary Construction Detail

SHEET-METAL BRACKET (1)

This bracket assignment is an introduction to sheet-metal and hand-tool operations. It may seem quite elementary, but from the standpoint of quality workmanship, a considerable amount of skill is required to construct the piece to meet production-line standards of quality.

The structure of the piece is simple enough so that full attention can be given to accuracy and careful workmanship in all phases of the fabrication. High-quality work is not an accident but comes about as the result of deliberate and concentrated effort. Figure 91 shows a pictorial view of the finished bracket piece.

This assignment includes making a sheet-metal layout, cutting, punching, and bending operations. In addition to these operations it is necessary to prepare the surface properly, deburr the edges, mount the swage nut, and mark the piece with name identification. All of this work must be done carefully.

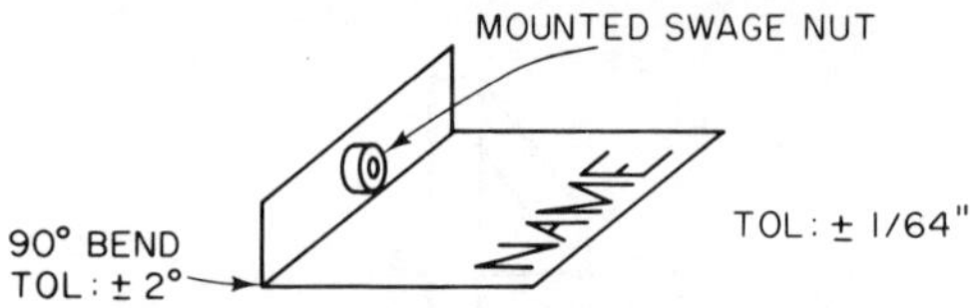

Figure 91 Sheet metal bracket

Working lines and *center lines* in a layout on sheet metal are used as a guide for subsequent fabrication operations. The accuracy and correctness of the workpiece is determined by the location of these lines. It is therefore necessary that all cut lines, hole centers, or reference lines be carefully placed in the proper position.

The *total* layout need not be completed before *all* machine- or hand-tool operations are performed. It depends upon the individual situation, and in some cases it is advantageous to do a part of the layout work at a later stage in the sequence. The exact *position* of certain critical lines (hole centers or other cut lines) can then sometimes be more easily determined. In some cases the dimensions might vary after certain operations (bending operations) have been performed, and errors can be eliminated by making the layout of these critical dimensions at a later stage. Also, certain fabrication operations can easily remove or dim out some of the layout lines, which could have been marked onto the piece later.

On sheet-metal work it is usually recommended that whenever possible, the cutting or punching operations be performed on the flat metal piece *before* the sides are bent up. It is easier to do these operations at this time. In view of this, the layout for these operations must be completed early in the sequence. As indicated above, the layout for certain other special operations can be made at a later time.

Sometimes the layout lines on the front need to be removed after the fabrication operations are completed. Because of this, it is well to remember to scratch all layout lines *very lightly* into the surface of the workpiece. Make the lines just heavy enough to *see and use,* especially on the front side of a front panel. In many cases the lines do not need to be made all the way across the piece, only in the area where the mark is to be used. Remember that quite often when a fine finish is applied the layout lines will show if not completely removed from the surface.

Removal of layout lines is best accomplished by using a waterproof garnet paper (400 to 600 grit) and flushing with plenty of water. Following this operation, the surface can also be caustic-etched in 15 to 20% sodium hydroxide solution, if the piece is made of aluminum.

Do not neglect to *double-check* all layout work. Much needless expense and waste of time can be prevented by putting forth the effort to check and double-check the layout work. It is foolish to cut a chassis piece 1/8″ short just because the dimensions were not checked or the location of the line was not determined to be correct.

The following items are suggested to be used as a guide in making a sheet-metal layout:

1. The preparation of *reference edges* on the blank stock. These can be edges that are 90° with respect to each other or at some other angle, depending upon the requirements. Measurements are made vertically from a horizontal edge and are made horizontally from a vertical edge. These edges are cut straight and true by the use of the squaring shear. The reference edges on a piece are shown in Figure 92.
2. The preparation of the *surface* for marking. Frequently, dirt or an oil film needs to be removed. In some cases a blue *metal-dye* can be painted on the surface to make the lines stand out. Because this dye must be removed at a later stage, its use is not recommended unless needed. The dye is very useful in covering up lines that are in error and allows the correct lines to stand out clearly.
3. The use of *wing dividers* for scribing circular lines or for transferring a measurement from the scale to the workpiece. This tool is used in a similar fashion to a compass, with one point of the dividers placed in a very lightly punched center-punch mark. Also, the exact distance between marks or lines can be accomplished by use of this tool. By setting the exact distance between the points with the scale, this distance can then be transferred directly to the workpiece. This is illustrated in Figure 93.
4. The proper use of the *scale*. The thickness of the scale (rule) and "sighting" problems often can cause layout errors. By using cau-

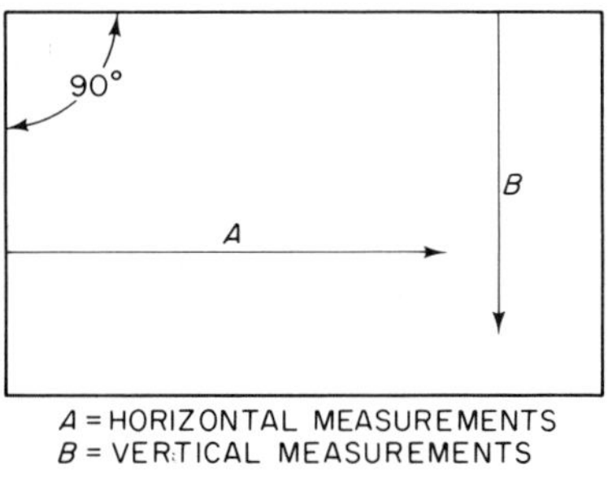

Figure 92 Reference edges on a workpiece

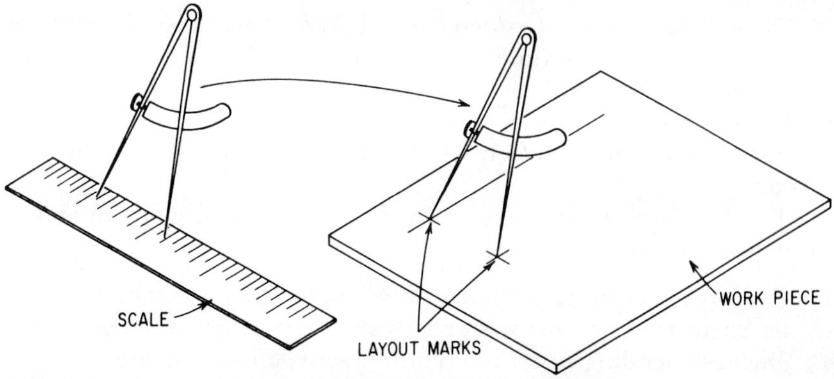

Figure 93 Use of wing dividers

tion and the proper technique, many errors can be prevented. One method of holding the scale is shown in Figure 94. As indicated in this sketch, the scale is held up on edge and the person sights *down* the scale lines. Using the scribe, a small mark is placed at the end of the proper scale line.

5. The use of the *combination square* for making the layout. A considerable degree of accuracy can be obtained by first setting the square to the dimension required. Then as the face of the head of the square is moved along one of the reference edges, an accurate line is scribed on the piece by holding the scribe at the end of the scale. This is shown in Figure 95. For each dimension on the layout the square is reset and the above procedure is repeated.
6. The determination of the sequence of making the layout and to what extent the layout will be completed before fabrication operations are begun. In some cases the location of a hole must be marked after certain bends are made.
7. The completion of the layout according to plans.

Purpose of Assignment

1. To gain experience in making a sheet-metal layout and using hand tools.
2. To become familiar with one type of swage-mounted nut.
3. To gain experience in reading a drawing.

Tools and Equipment

Combination square.

Scribe.

Steel rule.

Center punch and hammer.

Turret punch.

File.

Rivet squeezer.

C clamps.

Steel letter stamps.

Box-and-pan brake.

Squaring shear.

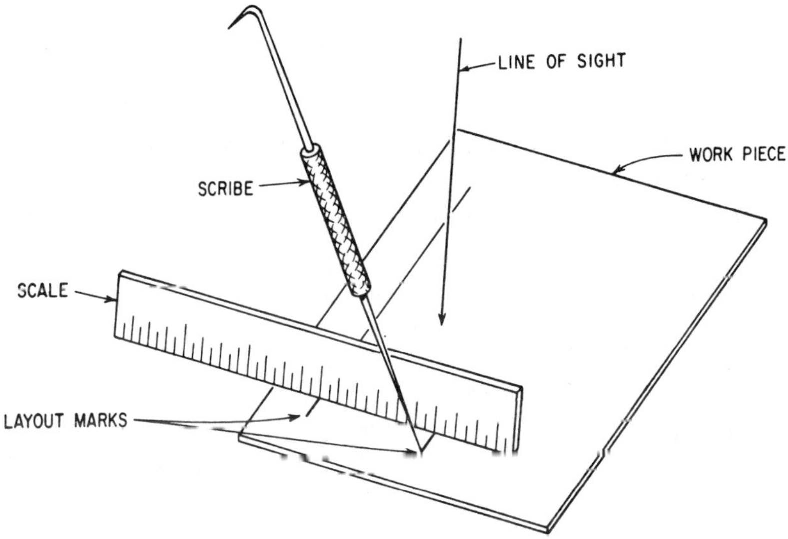

Figure 94 Proper use of the scale

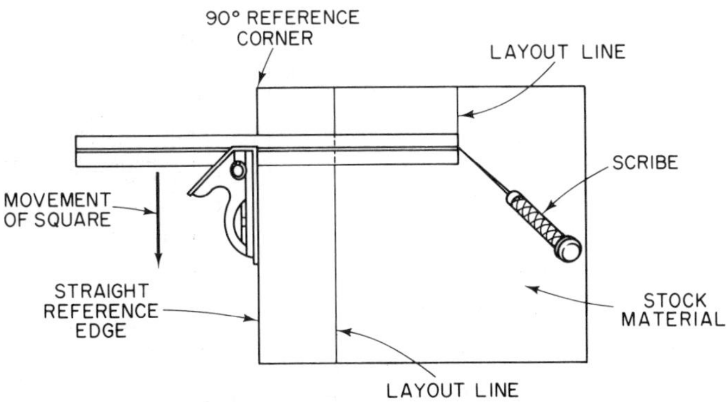

Figure 95 Making layout lines using a combination square

Materials

18-gauge aluminum, 2½ X 2½″.

1 swage nut, 6-32 machine screw thread (Standard Pressed Steel SN632-3 or equivalent).

320-, 400-, or 600-grit garnet paper.

Assignment

Make a sheet-metal bracket according to the directions given in Figure 96.

Directions

1. Remove scratches and prepare the surface of the metal piece to be used. Polish under water with a wetproof 320-, 400-, or 600-grit garnet paper.
2. Establish a 90° reference corner on blank stock (refer to the introduction).
3. Make the complete layout (refer to Figure 96).

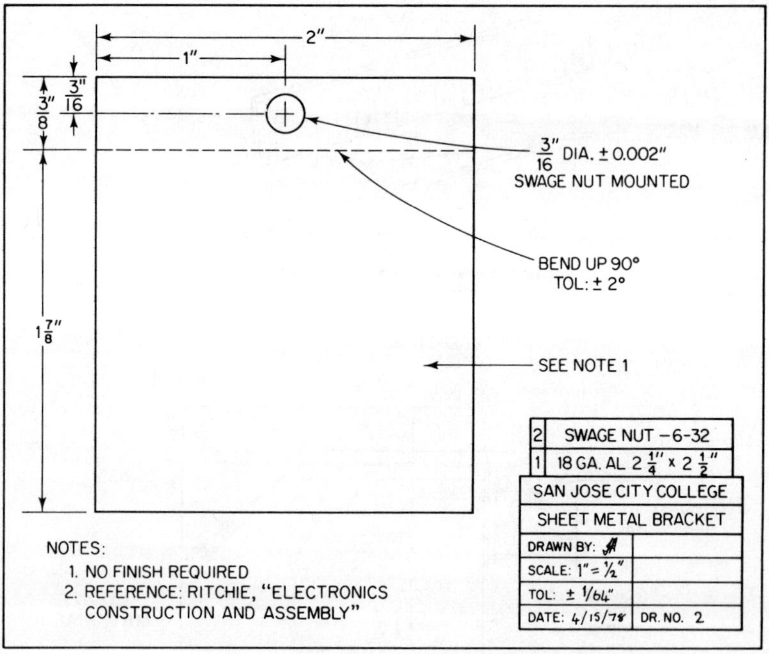

Figure 96 Sheet metal bracket

4. Perform the following hand-tool operations according to directions and instructions given by the instructor.
   a. Mark the center of the hole with the center punch.
   b. Punch the hole using the turret punch.
   c. Cut the outline using the squaring shear.
   d. Deburr all the edges and round the sharp corners very slightly.
   e. Mount the swage nut using the rivet squeezer (refer to the *Swage-Nut Mounting* information in Appendix B).
   f. Make the bend to 90°.
   g. Stamp your name on the piece (refer to the *Letter and Number Stamping* assignment).
5. When it is completed, turn in the piece to your instructor.

BEND ALLOWANCE FOR ALUMINUM (1)

Many factors affect the amount of bending compensation that needs to be considered in sheet-metal work. These factors include such things as (1) the properties of the material, (2) the characteristics of the forming tool, and (3) the working conditions of the actual bend.

In chassis-layout work, certain inside or outside measurement specifications often must be met. To meet these specifications, compensations must be made because of the bending characteristics of sheet metal. During the bend, the metal along the outside portion of the bend will undergo a *stretching,* and the metal along the inside portion of the bend will undergo a *compression.* The effect of these two items is illustrated in Figure 97 when bending #1100 or #3003 aluminum in a sharp bend.

As can be seen from a study of Figure 97, if the dimension is to be an inside measurement, a compensation of one-third the thickness of the metal (approximately) must be *added* for the *layout* dimension. If the dimension is to be an outside measurement, a compensation of two-thirds the thickness of the metal (approximately) must be *subtracted* for the layout dimension. In effect, in a bend as shown in Figure 97, the end position of the layout line is inside the side that is bent up. This occurs in both directions from the bend, and therefore a compensation must be made in both directions from the bend.

The procedure above, even though fairly accurate, is nevertheless a simplified approximation. It is a *good* approximation, however,

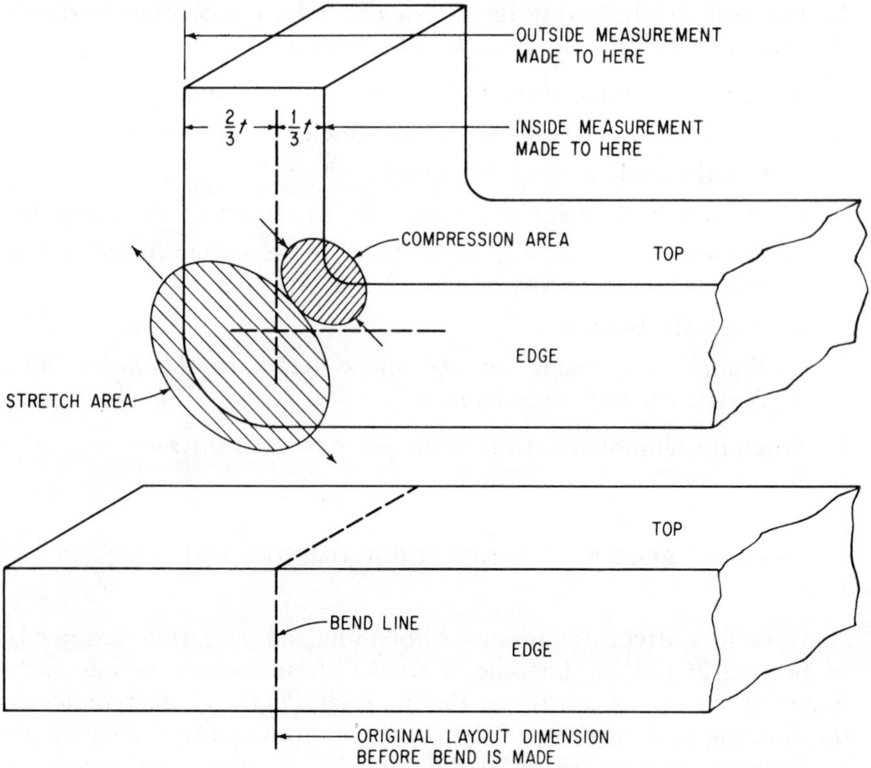

Figure 97 Bend allowance for aluminum

and will give results that are well within the range of the accuracy of the measuring instruments and techniques used in ordinary layout work. Because approximate values are dealt with in this system, both 16-gauge and 18-gauge aluminum can be considered to be 3/64″ thick. These are the thicknesses most commonly used in chassis construction.

Note that the dimensions in sheet-metal work can be *layout dimensions,* or dimensions that are the *result* of a certain layout. Also note that either of these can be *specified* or *calculated.* It follows also that there are *inside-to-inside* or *outside-to-outside* dimensions. As an example, there could be a *layout dimension* that is *specified* to give a required *inside-to-inside* measurement. Another example might be that from an *outside-to-outside* dimension requirement, a *layout dimension is calculated.*

Purpose of Assignment

1. To learn how to determine sheet-metal bend allowance, the reason for its application, and where it is applied.
2. To gain experience in measuring and working to a close tolerance.

Tools and Equipment

Combination square.
Scribe.
Steel rule.
File—flat, smooth mill file.
C clamps.
Steel letter stamps.
Box-and-pan brake.
Squaring shear.

Materials

2 pieces 18-gauge aluminum, 2½ X 2½" (#1100 or #3003).

Assignment

Bend the two angle pieces, shown in Figure 98, to the specifications given.

Directions

Note that both the *inside-to-inside* and *outside-to-outside* dimensions are *specified* and that the layout dimensions are to be *calculated*.

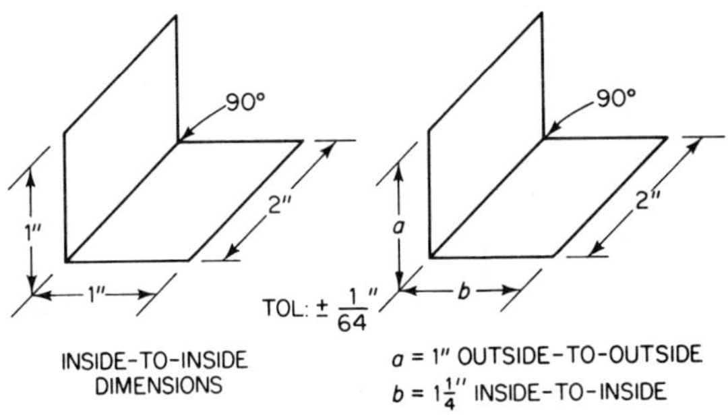

Figure 98 Specified dimensions for angle pieces

1. Calculate the layout dimensions required for construction of the pieces (refer to the introduction for bend-allowance information).
2. From a 90° reference corner, make a layout of the two pieces on the blank stock. Use a sharp, pointed scribe.
3. Cut the pieces to size and deburr all edges before making the 90° bends.
4. Letter-stamp your name on each piece and turn them in to your instructor.

BEND ALLOWANCE APPLICATION

During a bend, sheet metal will "stretch" and "compress" at the point where the bending is taking place. The magnitude of each of these effects will depend upon, for one thing, the characteristics of the metal. For example, aluminum is fairly soft and pliable and as a result will both stretch and compress at the bending point. This effect is shown in Figure 97. On the other hand, sheet steel is relatively tough and will squeeze together or compress very little. Therefore, in a bend with sheet steel, *stretching* takes place mainly on the outside portion of the bend (see Figure 99).

From a study of Figure 99 it can be seen that when using sheet steel, if the dimension is to be an *inside measurement,* the layout measurement will very closely equal the resultant inside measure. This means that no compensation for bend allowance is needed.

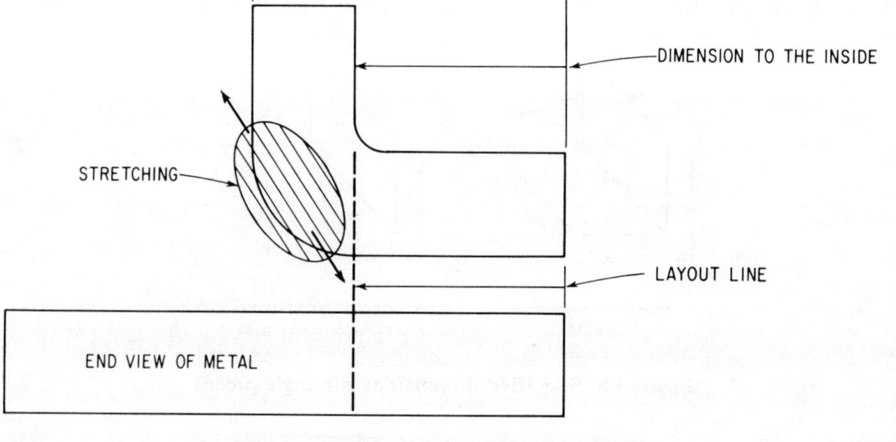

Figure 99 Bend allowance for sheet steel

Also, it can be seen that if the dimension is to be an *outside measure,* the thickness of the metal needs to be *subtracted* in locating the layout line. These results occur in both directions from the bend. Appropriate compensations, therefore, must be made in both directions.

When bending aluminum, as shown in Figure 97, the end position of the layout line is, in effect, *inside* the side that is bent up. This takes place in both directions from the bend, and therefore a compensation must be made in both directions. If the dimension is to be an inside measurement, a compensation of one-third the thickness of the metal (approximately) must be *added* for the layout dimension. If the dimension is to be an outside measurement, a compensation of two-thirds the thickness of the metal (approximately) must be *subtracted* for the layout dimension.

The procedure described above, even though fairly accurate, is nevertheless a simplified approximation. It is a *good* approximation, however, and will give results that are well within the accuracy of the measuring instruments and techniques used in an ordinary small shop.

As an example, as shown in Figure 100, there is a metal cover that needs to fit into a box that has 3″ inside clearance. For calculations both 16- or 18-gauge aluminum would be considered to have a thickness of 3/64″. Then

$$1/3t = 1/64''$$

$$2/3t = 2/64'' = 1/32''$$

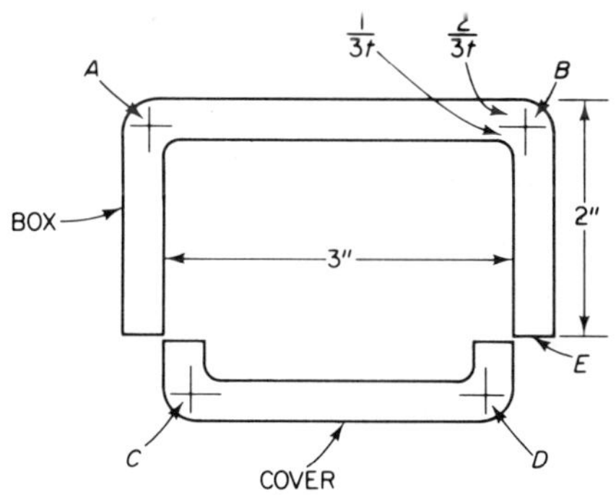

FOR BENDING THE BOX:

LAYOUT DIMENSION *A* TO *B* $= 3'' + \frac{1}{64}'' + \frac{1}{64}'' = 3\frac{1}{32}''$

LAYOUT DIMENSION *B* TO *E* $= 2'' - \frac{1}{32}'' = 1\frac{31}{32}''$

FOR BENDING THE COVER:

LAYOUT DIMENSION *C* TO *D* $= 3'' - \frac{1}{32}'' - \frac{1}{32}'' = 2\frac{15}{16}''$

Figure 100 Bend allowance illustration

Purpose of Assignment

1. To become acquainted with the bending characteristics of metals.
2. To learn how to determine sheet-metal bend allowance for aluminum.

Tools and Equipment

Combination square.

Scribe.

Steel rule.

File—flat, smooth mill file.

C clamps.

Steel letter stamps.

Box-and-pan brake.

Squaring shear.

Materials

2 pieces 16-gauge aluminum, 2 X 6" (#1100 or #3003).

Assignment

From the information given in Figure 101, construct two U-shaped pieces to fit together. These are to fit one inside the other and the layout dimensions must be determined.

Directions

1. Refer to the introduction for bend-allowance information.
2. From the specifications given in Figure 101, determine the *layout dimensions* of the U-shaped pieces. These pieces are supposed to fit one inside the other. Consider 16-gauge aluminum to be 3/64" thick.
3. Notice that the requirements are given in terms of *inside-to-inside* or *outside-to-outside* dimensions. In this case the layout dimensions are to be *calculated.*
4. Fill out the dimension information in Figure 102 for the layout of the pieces.
5. From a 90° reference corner, make a layout of the two pieces on blank stock. Use 16-gauge aluminum and work to a tolerance of ±1/64".

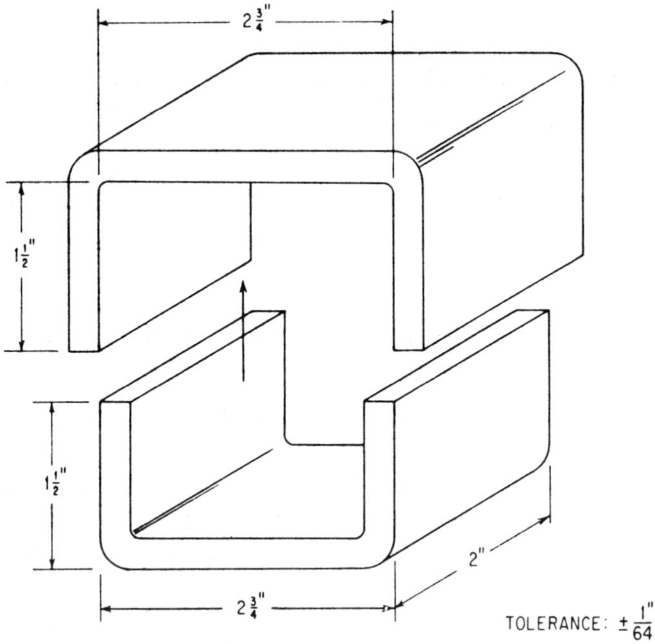

Figure 101 U-shaped pieces

BEND ALLOWANCE APPLICATION
WORK SHEET

FILL IN THE REQUIRED DIMENSION INFORMATION FOR CONSTRUCTION OF THE U-SHAPED BEND PIECES.

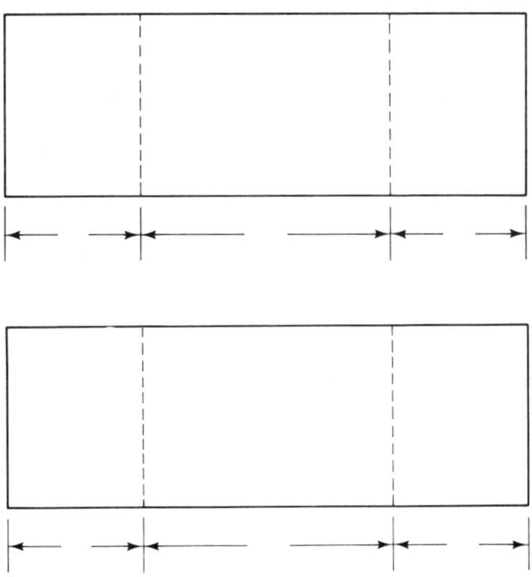

Figure 102 Layout calculations for U-shaped bends

6. Cut to size and deburr all edges before making the 90° bends.
7. After making the bends, check for a perfect fit of the two pieces together.
8. Letter-stamp your name on each piece and turn them in to your instructor.

PROBLEMS FOR BEND ALLOWANCE

Refer to the information given in *Bend Allowance for Aluminum* and *Bend-Allowance Application.*

1. What is the layout dimension x for the bottom cover piece designed to fit the box shown in Figure 103? The piece must have lips on the front and back edges, which are to fit inside the

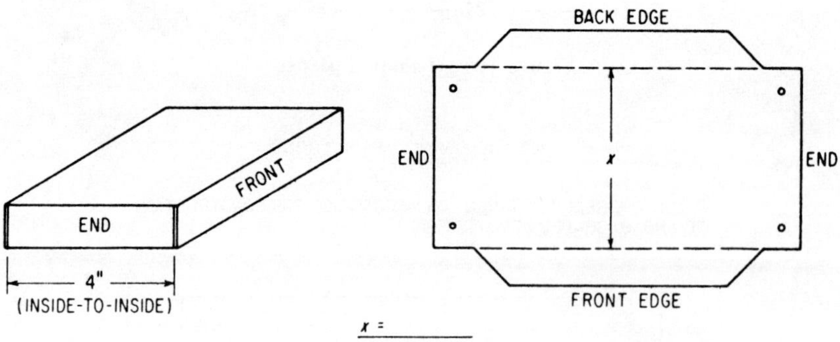

Figure 103 Inside-to-inside bend allowance

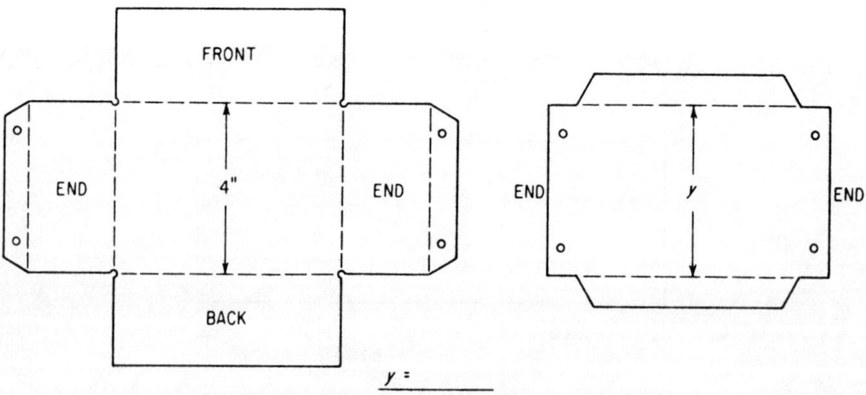

Figure 104 Inside-to-inside layout dimensions

front and back edges of the box. The dimension given for the box is an inside-to-inside dimension. Consider the aluminum to be 3/64″ thick for all problems.

2. For the same type of box as shown in Figure 103, and from the layout of the box illustrated in Figure 104, what is the bottom-piece layout dimension *y*?
3. For the same type of box as shown above, what are the dimensions *A* and *B* if the box is to be 4″ outside to outside? (Refer to Figure 105.)

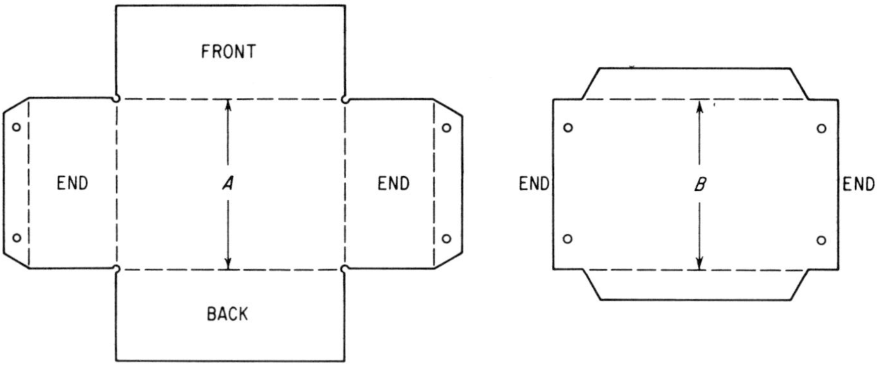

A = ________

B = ________

Figure 105 Outside-to-outside bend allowance

SUBCHASSIS CONSTRUCTION (1)

This job incorporates a number of hand-tool or machine-tool operations. Some of these operations are *interdependent* and require a small amount of forethought on the part of the person constructing this piece. If the wrong order is followed in doing some of these operations, the whole job is made considerably more difficult. Be careful and *think through the order that should be followed.* In some cases, it becomes impossible or nearly impossible to complete certain operations when the proper order is not followed. As an *example,* the operations on a simple construction should be in the following order:

Step 1: Cut to size.

Step 2: Bend into shape.

Step 3: Paint the piece.

The following is a *basic sequence* that should be followed in constructing a simple sheet-metal piece:

1. Cut sheet stock and allow 3/8 to 1/2″ for squaring and trimming.
2. Establish straight reference edges at 90°.
3. Lay out work on blank.
   a. Cut lines.
   b. Hole center locations.
   c. Bend lines.
4. Perform hand- and machine-tool operations in the *proper order.* (Drill or punch holes, make cutouts, cut outlines, and make bends. Fit other special operations somewhere in this sequence.)
5. Spot-weld, if applicable.
6. Prepare and finish the surface.

Commonly used hand tools such as the square, scribe, scale, center punch, tin shears, and the file are introduced in this assignment. The instructor will also demonstrate the use of certain machine tools, which then will give the student a few ideas of what tools are involved in ordinary chassis construction. More ideas, of course, will come later as other assignments are given.

Purpose of Assignment

1. To gain experience in reading a layout drawing.
2. To gain experience in the use of basic hand tools and machines.
3. To develop the ability to look ahead and *plan a sequence* of construction operations.
4. To develop the ability to work carefully and to close tolerances.

Tools and Equipment

Various hand tools.

Squaring shear.

Corner notcher.

Brake.

Tin shears.

Drill press.

Dimpler or standoff depressor.

Turret punch.

Materials

$5\frac{1}{8} \times 6\frac{1}{4}''$ 16-gauge aluminum.

Assignment

Using 16-gauge aluminum, construct a subchassis from the information given in the drawing of Figure 106. Instructions are given in the Directions.

Directions

1. Follow the basic sequence for construction of a simple sheet-metal piece as given in the introduction.
2. Establish an accurate 90° reference corner and measure all layout dimensions from the edges of this corner. Work to close tolerances

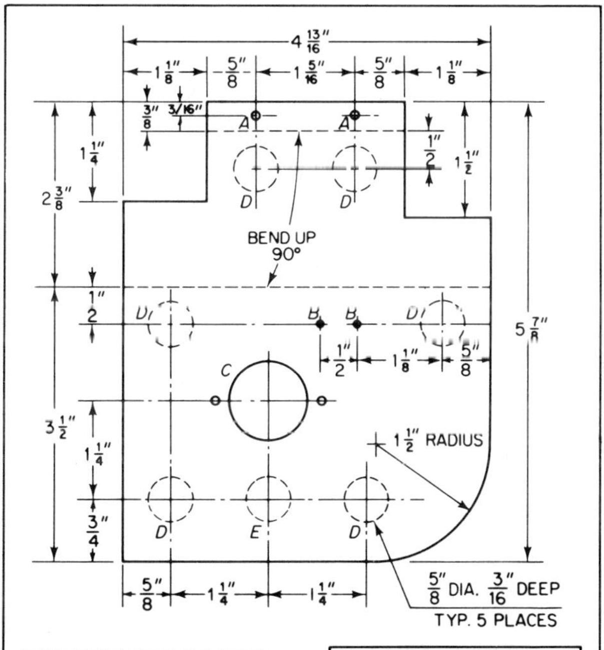

DATA CHART	
A	ROTEX PUNCH D
B	0.081" DIAM. ± 0.003"
C	ROTEX PUNCH P
D	DEPRESS DOWN
E	DEPRESS UP

SUBCHASSIS	
DATE: 2/5/78	MATERIAL: AL.
DRAWN BY: JH	GAUGE: 16
SCALE: 1" = 1"	FINISH: NONE
TOL: ± 1/64	
	DR. NO.: 1

Figure 106 Subchassis construction detail

on the layout, cutting, drilling, and the bending. No bend allowances are needed in the construction of this piece. Work carefully and accurately. Think ahead as you work and *plan a sequence* of the operations to be performed.

3. Deburr *all* edges.
4. Stamp your name on the piece before turning it in to your instructor.

CORNER SAMPLE NO. 1 (1)

The instrument chassis or cabinet might vary in its design, but it is likely that it could be a wraparound type and include a corner such as that illustrated in Figure 107. The cabinet includes a wraparound piece (forming the top, sides, and bottom), front and back face panels, and a shelf (attached to the front panel) for component mounting. In some cases, the components are all mounted on the front panel and are removed with this panel when access to them is desired.

This type of chassis is used because it makes an electronics instrument package that is neat and quite versatile as to the type of circuit that can be incorporated within it. Further, this type of chassis is adaptable to the small shop where a limited number of tools is available.

The wraparound feature alleviates some of the problems of chassis construction, but it also introduces problems that are peculiar to it. Some of these problems are as follows:

1. The method of joining the ends of the wraparound piece either at a corner (or corners), or in the middle of the bottom.
2. The construction of the sides the same height, or the top and bottom the same length.

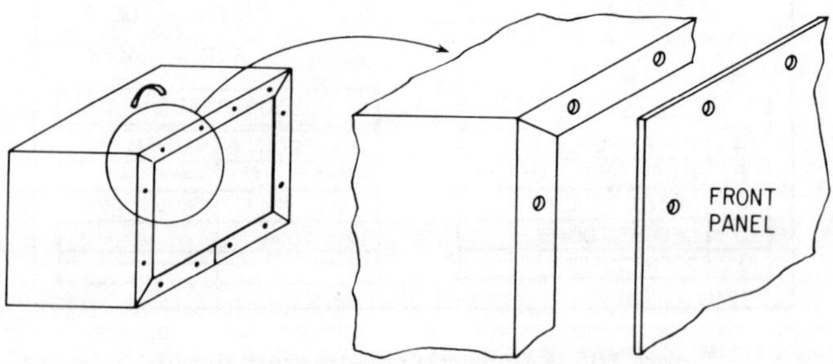

Figure 107 Wraparound chassis corner

3. The bending of the front and back lips all at the same time with a brake that will handle only a relatively short bend.
4. The bending of the four corners of the wraparound piece without interference of the bending machine.
5. The attachment of the component shelf to the front panel.

These problems can all be solved using caution and good judgment. It is suggested that the joint in the wraparound piece be made in the middle of the bottom by spot-welding a strap across the joint on the inside as shown in Figure 108.

Purpose of Assignment

1. To introduce one type of commonly used cabinet or chassis corner.
2. To gain experience in the use of tools and machines used in sheet-metal work.
3. To develop the habits of careful and accurate workmanship.
4. To learn about chemical conversion finishes.

Tools and Equipment

Various hand tools.

Box-and-pan brake and press brake.

Corner notcher.

Drill press or turret punch.

Squaring shear.

#56 drill.

Chemical tanks as required.

Materials

18-gauge sheet aluminum, 2 × 5″.

20-gauge sheet steel, 2 × 5″.

Chemicals for an alodine coating (refer to the *Alodine Processing* assignment).

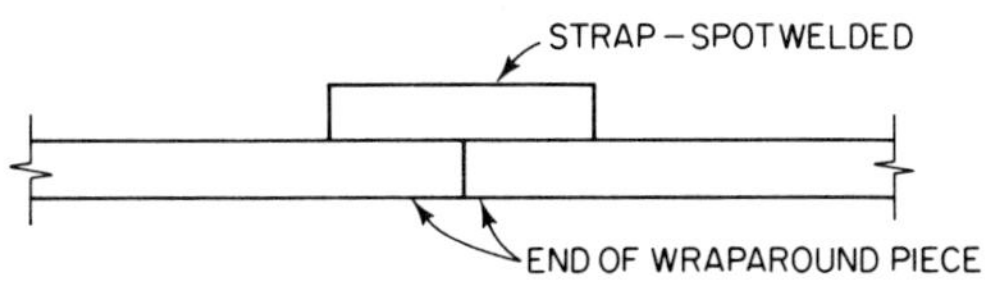

Figure 108 Joining wraparound piece

Assignment

Using the aluminum and steel pieces, construct two sample corners from the layout information given in Figure 109.

Directions

1. Using the aluminum piece, establish reference edges and lay out the bend and cut lines, and the center for holes.

CAUTION: The *depth of the notch* and the relative *position of the bend line* for the lips is critical. Essentially, if a tight bend is made and the vertex of the notch is inside past the bend line, there will be a gap in the corner between the lips. If the vertex of the notch is outside the bend line, the lips will hit together and prevent the sides from being bent up to 90°.

---

2. Perform other necessary operations in the *proper sequence.* Think through the sequence and the interdependence of the operations that need to be performed. The order of making the bends is indicated in Figure 109.
3. Alodine the aluminum corner piece when it is finished. Hold it in the solutions with a small wire through the hole in the corner of the piece (refer to the *Alodine Processing* assignment).
4. Construct another sample corner of the same type using the sheet iron if required by your instructor.
5. Stamp your name on the corners and turn them in (use the steel letter stamps and refer to the *Letter and Number Stamping* assignment).

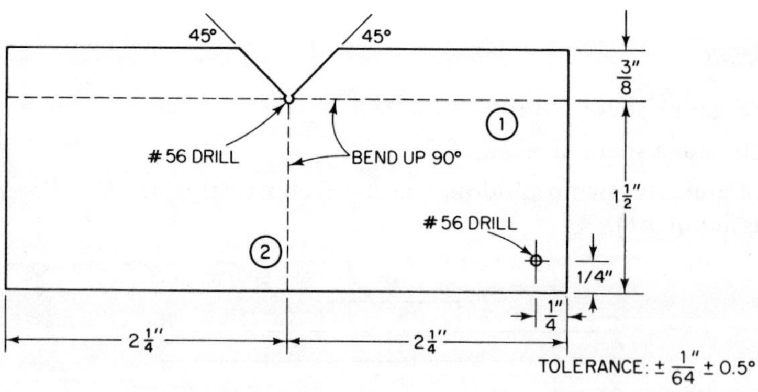

Figure 109 Layout of Corner Sample No. 1

The corner illustrated in this assignment is often used to form the open-type box chassis similar to that shown in Figure 110. *F* is the top of the chassis.

This corner includes the addition of a *lip*, which is used for the attachment of a cover or bottom plate. Controls could be mounted on front side *A* or back side *D*, the major components mounted on top, and small parts mounted underneath on the inside. A bottom cover, with feet, is usually attached from the underneath side. The spot-weld tabs *B* are used for securing the corners. Inset of the spot-weld tabs is accomplished in the manner suggested below.

In fabricating this corner there might be three separate compensations made in order to have it come out right. First, a compensation allowance for the height of sides *A* and *D* is made so that these sides will match the height of sides *C* and *E*. This is shown in Figure 111. Second, in some cases sides *A* and *D* are extended an additional thickness of the metal to hide the edge of the bottom cover piece. This is illustrated in Figure 112. Third, when an attachment lip is included, like the one on the outside edge of side *C*, the total resulting height of that side will be increased. This must be counted when calculating the relative heights of each side. These calculations or compensations for the sides are made keeping in mind that the height you would be working *from* or working *toward* is predictable using sheet-metal bend-allowance information.

In the construction of this corner, which incorporates use of the spot-weld tab (see Figures 111 and 113), special consideration must

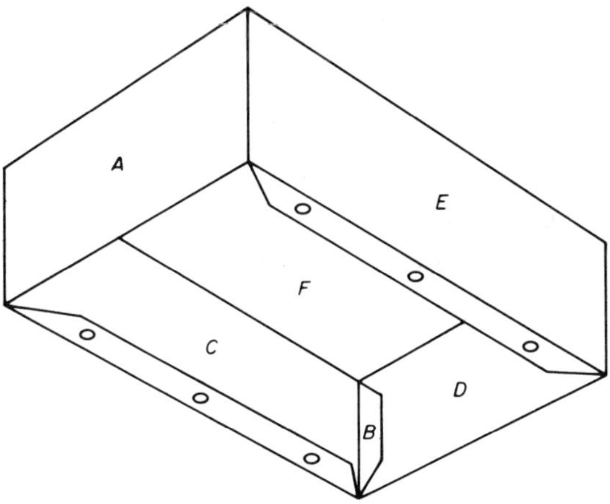

Figure 110 Box chassis

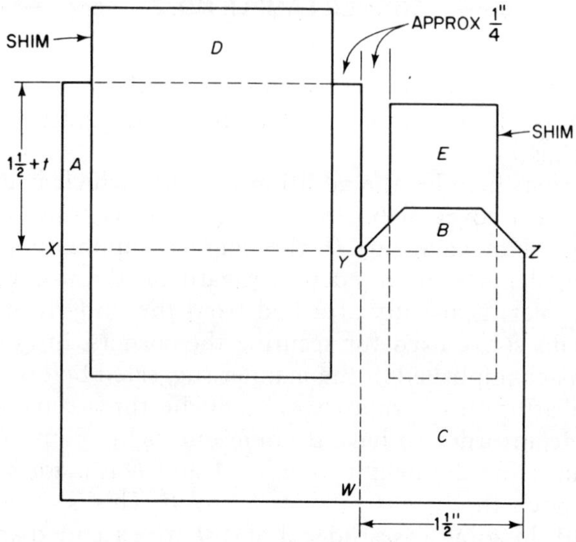

Figure 111 Placement of shims for tab inset

be given to the inset of tab *B* with respect to end piece *A*. If this is not done, tab *B* will hit directly against end piece *A* when bend *YW* is made. One of two methods can be followed to prevent this.

First, in the layout, in reference to Figures 111 and 113, bend line *YZ* can be shifted inward by one thickness of the metal from line *XY*. The bends (on line *YZ* and line *XY*) are then made separately and an inset of tab *B* is accomplished.

Second, in the layout, line *XYZ* is a straight line and a special shim arrangement of bending tab *B* and end piece *A* can be used. This method is quite simple, accomplishes an automatic inset of tab *B*, and is described as follows:

1. In the layout of the piece for the corner, *add* the amount of one thickness of the metal to the height of end piece *A*. During the bending operation, the actual bending place of end piece *A* is one thickness of the metal outward from line *XY*. A compensation must be made for this so that end piece *A* will not end up short.

NOTE: Refer to Figure 111. It is assumed at this point that the layout work has been completed, the outline cut, relief hole drilled, and the edges deburred.

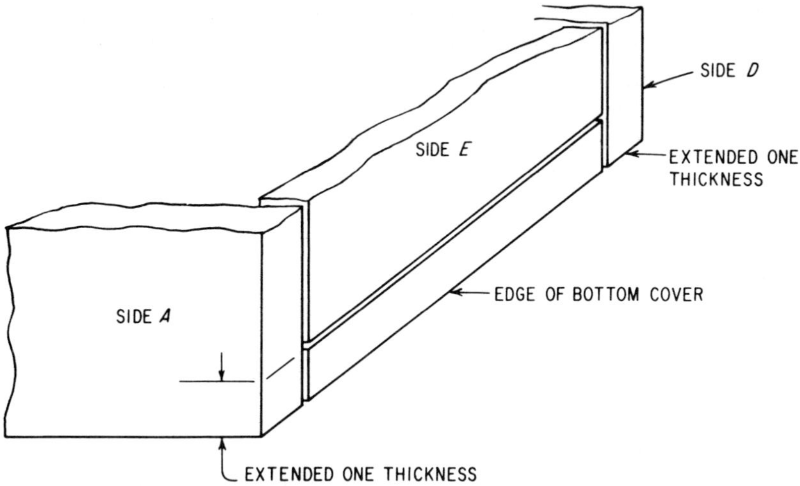

Figure 112 Enlarged view of box-chassis side

2. Obtain two shims of scrap metal the same thickness as the metal used for the corner.
   a. One shim is approximately 3″ long and 2½″ wide.
   b. One shim is approximately 3″ long and 1¼″ wide.
3. Place the corner piece in the brake to be bent on line *XZ*.
4. Place one shim of scrap metal *(E) under* tab *B* as shown in Figure 111.
5. Place one shim of scrap metal *(D) over* end piece *A*, as shown in Figure 111.
6. Bend tab *B* and end piece *A* up 90°. This will accomplish an automatic inset of tab *B*.

NOTE: This simplified corner, although sometimes used in actual construction, has been given mainly for the purpose of illustrating this particular method of bending the ends of a box and at the same time obtaining an inset of the spot-weld tabs.

An instrument case can also be formed by the use of this type of corner. In this case (refer to Figure 110) lips for the attachment of the front panel are added to sides *A* and *D*. The length, width, and height are adjusted as needed for the circuit components or other physical requirements.

Again, as noted above, an allowance for correcting the relative heights of the sides is required in the layout. This will be true whenever the special method of bending the tabs and sides is used, or when the attachment lips are included.

When using this type of construction, the main problems encountered are as follows:

1. Accomplishing an inset of the spot-weld tabs with respect to the side pieces.
2. Making corrections in the layout to accomplish equal heights of the sides.
3. Following a sequential order in making the bends.
4. Mounting the captive or swage nuts at the proper place in the sequence of operations.

The swage nuts are mounted according to the instructions for the particular nuts used. These nuts are used whenever they are located on the inside where they cannot be held while tightening. In instrument chassis, they are extensively used in lieu of sheet-metal screws. For mounting these nuts, refer to the information in Appendix B on *Swage-Nut Mounting*.

Purpose of Assignment

1. To introduce one type of commonly used cabinet or box-chassis corner.
2. To gain experience in spot welding.
3. To gain experience in bending techniques, including the application of bend allowances.
4. To make application of end-piece allowance and spot-weld tab inset.
5. To become acquainted with captive or swage-mounted chassis nuts.

Tools and Equipment

Various hand tools.

Brake.

Corner notcher.

Spot welder.

Drill press or turret punch.

Squaring shear.

Materials

18-gauge sheet aluminum, 4 X 5″.

One 6–32 swage nut (Standard Pressed Steel SN632–3 or equivalent).

Scrap pieces of 18-gauge aluminum.

Assignment

Construct the corner sample from the information given in Figure 113.

Directions

1. Predict the resultant height of side *C* from the information given in Figure 113. As a hint for this, the calculations would start from the basic 1½″ layout dimension given for side *C*. Add to this the resulting effect of including the attachment lip bent at 90° (refer to the introduction of the *Bend Allowance for Aluminum* assignment).
2. Determine dimension *H*, which must include allowances so that the resultant height of side *A* will be the same as side *C*. Base your calculations on the instructions given for the special method of bending tab *B* and side *A* as described in the introduction and

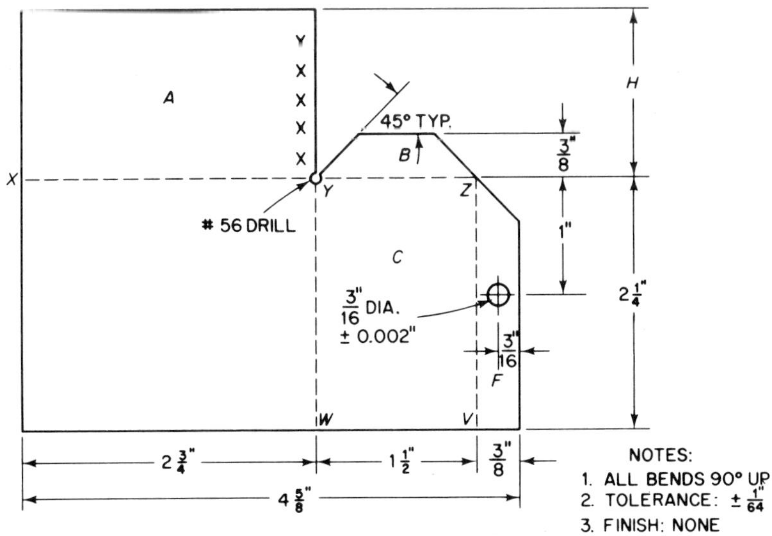

Figure 113 Layout of Corner Sample No. 2

also the fact that the attachment lips on side *C* will cause an increase in the height of that side. Two factors must be included in order to determine dimension *H*.

3. Make your layout from 90° reference edges, and punch or drill the relief hole and the hole for the swage nut.
4. Cut the outline and deburr the edges to be smooth.
5. Mount the swage nut in *F* and then bend line *ZV*.
6. Continue the operations in *the proper order*.
7. Spot-weld tab *B* to side *A*. Refer to the *Spot Welding* assignment.
8. Stamp your name on the corner and turn it in to your instructor.

CORNER SAMPLE NO. 3

This corner is just another version of the two preceding corners. The allowances, techniques, and procedures that apply in the construction of those corners also will apply in the construction of this corner. Figure 114 shows a view of the chassis made with this type of corner.

The problems encountered in making this type of corner are as follows (refer to Figure 115):

1. The determination of dimension *P*. It is figured from the resultant outside dimension of tab *B* after it is bent and the inside measurement of lip *D* after it is bent. Bend-allowance calculations need to be considered. Lip *D* fits into cutout portion *P*.
2. Extension of side *C* 3/64" in the direction of the arrow. This allows the final form of the corner to be filled in more and covers the end of side *A* after it is bent up.

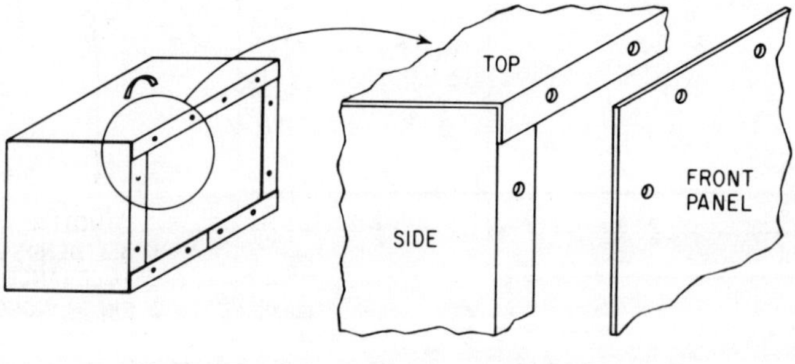

Figure 114 Chassis corner view

3. Position of the relief hole. It is best located exactly at the point where the bend lines cross.
4. An addition of 3/64″ made in the layout of side *C*. This is done when the special method of bending up tab *B* and side *C* is used. This method is described in *Corner Sample No. 2* and requires that there be an addition to the layout length of side *C*. The amount added is equal to one thickness of the metal or for 16-gauge aluminum, this would be 3/64″.
5. The order of operations is important. In addition to other operations, lips *D* and *E* should be bent up first, the swage nuts mounted, side *C* and tab *B* bent up next, and then side *A* bent up last.

With this corner, good results can be obtained by leaving off the spot-weld tab *B*. In this case (shown in Figure 115) dimension *S*

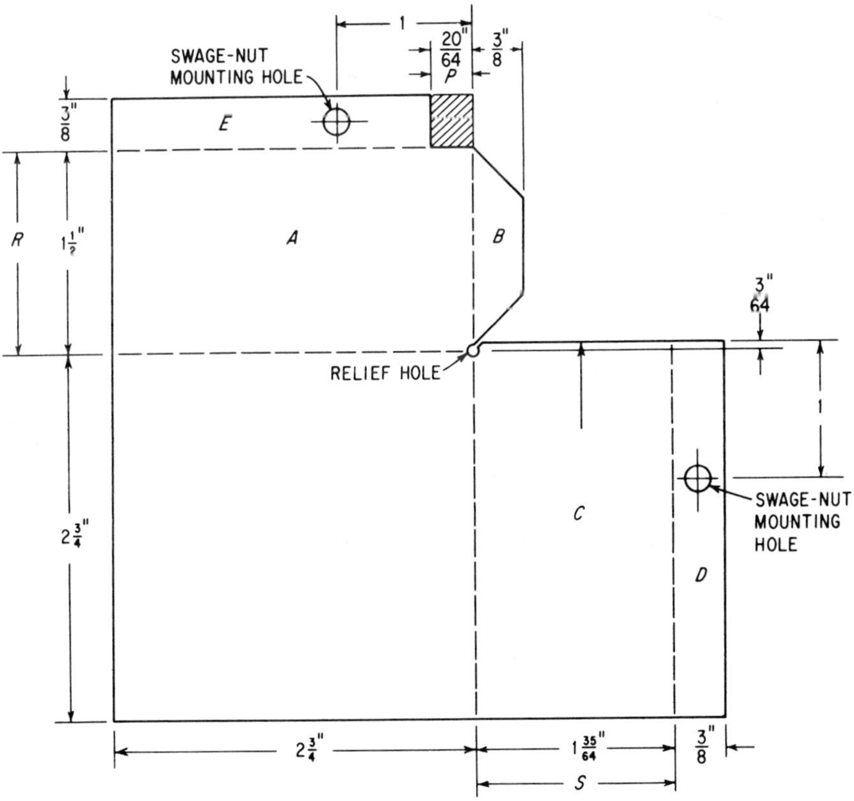

Figure 115 Layout of Corner Sample No. 3

should equal dimension *R* and as indicated above, the layout of side *C* is extended in the direction of the arrow to trim up the corner.

Purpose of Assignment

1. To introduce one type of commonly used cabinet or box-chassis corner.
2. To gain experience in spot welding.
3. To gain experience in bending techniques, including the application of bend allowances.
4. To make application of the spot welder and the use of captive or swage-mounted nuts.
5. To gain experience in making the layout of corners that must fold up and fit together.

Tools and Equipment

Various hand tools.

Brake.

Corner notcher.

Drill press or turret punch.

Spot welder.

Squaring shear.

Materials

16-gauge sheet aluminum, 5 × 5″.

One 6–32 swage nut (Standard Pressed Steel SN632–3 or equivalent).

Scrap pieces of 16-gauge aluminum.

Assignment

Construct the corner from the information given in Figure 115.

Directions

1. Make the layout from 90° reference corners.
2. Determine the *proper sequence* for *all* the operations *before* beginning the actual construction.
3. Complete the machine- and hand-tool operations.
4. When the piece is completed, stamp your name on it and turn it in to your instructor.

5

Construction of the Typical Chassis

METER BOX CONSTRUCTION

The meter box is a very important but simple type of construction. It is often used to house a meter circuit or some other similar circuit where the components can all be mounted directly on the front panel or face plate. In this case the front panel is the frame and the box is simply a cover.

This assignment incorporates some of the ideas and techniques previously discussed. It is intended that, as a separate assignment, a face plate (not shown in Figure 116) be constructed to match the box.

In the construction of this box, special attention must be given to ensure that the four sides of the box will all be the same height and that the spot-weld tabs will clear the end pieces when the sides are bent up. These two problems have been taken care of by the recommendation of a *special shim arrangement* for bending the tabs and end pieces and by the inclusion of a *necessary end-piece allowance* in the layout drawing. The instructions for this method of bending are given in the Directions.

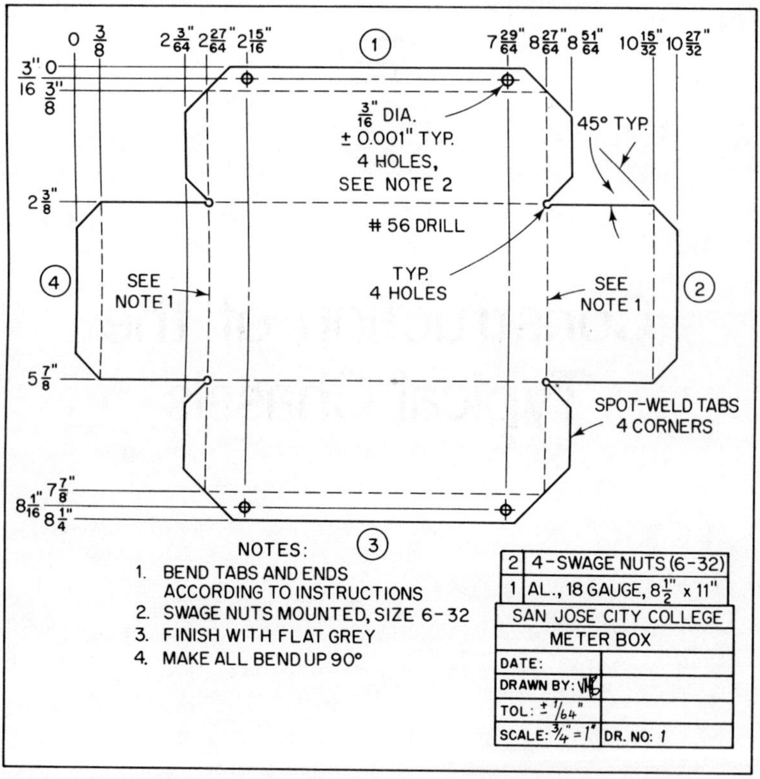

Figure 116 Meter-box construction detail

Purpose of Assignment

1. To construct a basic type of chassis.
2. To gain experience in the refinement of sheet-metal bending techniques.
3. To gain experience in the use of tools in accomplishing careful and accurate workmanship.

Tools and Equipment

Hand tools.
Squaring shear.
Brake.
Corner notcher.
Turret-type chassis punch.
Spot welder.

Materials

18-gauge aluminum sheet, 9 × 11″.

Four swage-mounted nuts, size 6–32 (Standard Pressed Steel SN632–3 or equivalent).

Paint.

Assignment

Using 18-gauge aluminum, construct a meter box according to the instructions given in Figure 116.

Directions

1. Establish an accurate 90° reference corner on the blank stock and make all layout measurements from there.
2. Establish a *sequence* for the operations to be performed. Work to close tolerances on the layout, cutting, drilling or punching, and the bending. All necessary allowances have been made and included in the drawing.
3. Install the swage-mounted nuts carefully.
4. Follow the special instructions given below for bending the spot-weld tabs and the end pieces. Provisions and allowances have been made in Figure 116 so that this method will automatically accomplish the exact tab inset and equality in the height of the sides.

NOTE: Refer to Figure 111. It is assumed at this point that the layout work has been completed, the outline cut, holes drilled or punched, side lips 1, 2, 3, and 4 have been bent up 90°, the edges deburred, and the swage-mounted nuts have been installed.

---

a. Obtain three shims of scrap metal the same thickness as the metal used for the box. Two shims should be approximately 3″ long and as wide as the sides of the box less 1/2″ and the third shim should be approximately 3″ long and as wide as the end piece less 1/2″.
b. Place the box in the brake to be bent on line *RZ*.
c. Place two shims of scrap metal *under* tabs *A* and *C*, respectively, as shown in Figure 117.
d. Place one shim of scrap metal *over* end piece *B* as shown in Figure 117.

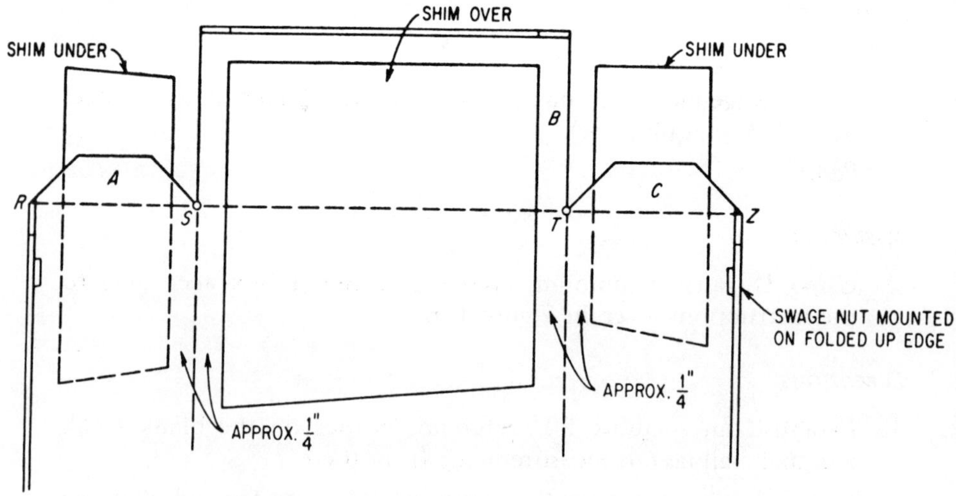

Figure 117 Shim arrangement for tab inset

e. Bend tabs *A* and *C* and end piece *B* up 90°. This will accomplish an automatic inset of tabs *A* and *C*.

f. Repeat with a similar operation on the opposite end of the box.

5. Make any slight adjustments or corrections to the box after the two sides are bent up.
6. Spot-weld the tabs to the sides of the box. Refer to *Spot Welding*.
7. Stamp your name on the inside of the box with the letter stamps. Refer to *Letter and Number Stamping*.
8. Spray-paint the box. Refer to *Spray Painting*. When completed, turn the box in to your instructor.

SPOT WELDING

(continued from previous assignment)

Sheet-metal pieces can be very simply and firmly joined together by the use of the spot welder. Essentially, the metal pieces to be joined are held tightly together between two welding tips and a current of considerable magnitude is passed through the tips and the work. This current is of a very short duration and must be sufficient to cause the work pieces to fuse together. The current required to weld sheet steel (depending upon the amount of time) will usually be less than the current required to weld aluminum.

The spot welder operates on the principle that heat develops at points of electrical resistance. This is borne out by the relationship P (heat) $= I^2 R$. In a spot-welder setup, the point of highest resistance is between the work pieces.

It is suggested that a flat tip be used against the *outside* surface as shown in Figure 118. This prevents the formation of a deep spot weld and allows the box to be painted without the spot showing. A small tip will mar the surface. The upper electrode tip diameter should approximately equal $0.1 + 2t$ (where t = thickness of metal).

There are many different types of spot welders, including manually operated, manually timed, air-operated, automatically timed, and motor-driven welders.

Purpose of Assignment

1. To become acquainted with the function and use of the spot welder.
2. To develop the habit of following directions and planning an orderly sequence of operations.

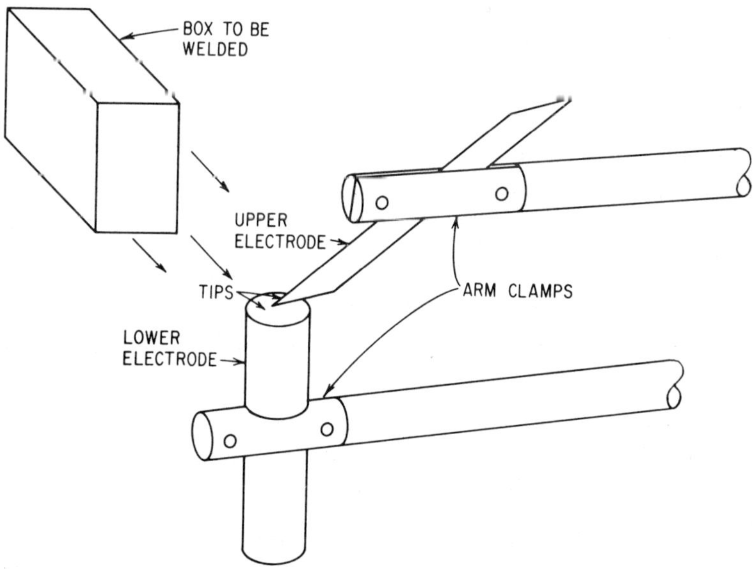

Figure 118 Spot-welder electrodes

Tools and Equipment

Capacitor discharge spot welder or a welder with adequate current output for spot-welding aluminum (with instruction manual).

Smooth file.

Allen wrenches.

Adjustable end wrench.

C clamps, as needed.

Goggles for eye protection.

Materials

Scrap pieces of aluminum.

Item to be spot-welded.

Emery cloth.

Assignment

1. Set up the spot welder according to a logical, orderly sequence. The sequence given in the Directions may be followed.
2. Complete a number of practice welds and then complete the welding on a workpiece as in Figure 119.

Directions

The following sequence is suggested as a logical method of setting up an air-operated, automatically timed spot welder. The adjustments are set and the welder is made ready to use *before* the power is turned on.

1. Set up welder arms and electrodes to fit the work.
    a. Clean the electrodes to make sure they make good contact in the arm clamps.
    b. Adjust the electrodes in the arm clamps to make sure they have the proper *opening clearance* and *closing pressure.* This may vary depending upon the welder.
2. Adjust timer controls, heat controls, and transformer taps. Start with heat setting on the *low side* and then, after experimenting, increase if necessary.
3. Turn on *cooling water* and *operating air.*

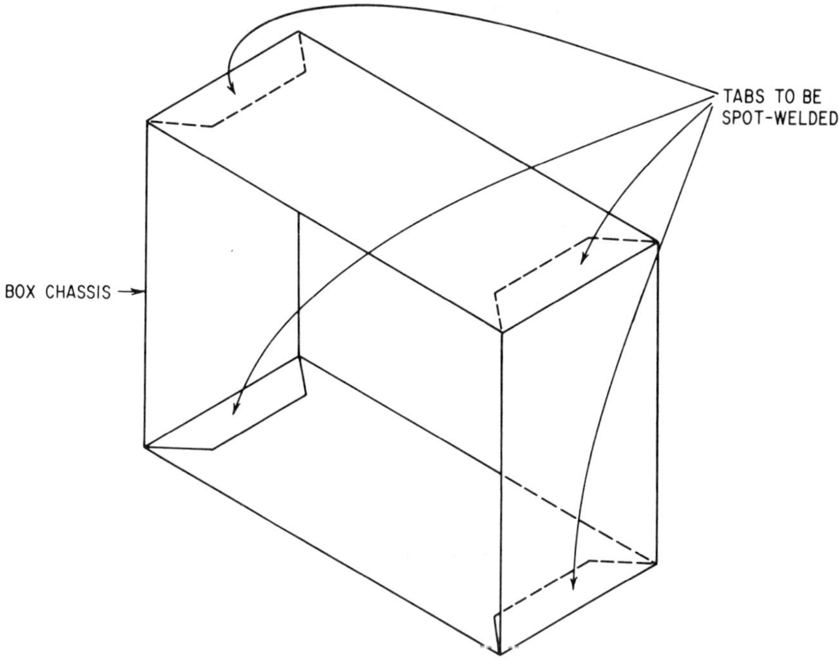

Figure 119 Chassis to be spot-welded

a. *Double check* for water flow through the welder.
b. Check the air pressure.

4. Turn on electric power. Be sure that the "no weld" switch is in "off" position until actually ready for welding. Put on safety goggles.
5. Check the welder.
   a. Try a few *practice welds* on scrap pieces to check the operation of the welder and the control settings. If the heat settings are low, the welds will not fuse properly; and if the heat is too high, the welds will tend to burn or explode.
   b. Keep the welding tips *clean*. Check them often. Use the file or the emery cloth.
6. Tack the workpieces with *one spot* and check for proper relative position of pieces. Sometimes the pieces can be separated and respotted if necessary. Separate by *twisting* and not by pulling apart, or *shift* one of the pieces sideways by tapping with a hammer.
7. Finish welding the work item as required.

—/\/\/\— METER BOX FACE PLATE —/\/\/\—

(continued from previous assignment)

All parts of a circuit are often mounted directly on the front panel or on the face plate. The rest of the chassis, then, is simply a cover. This is true of the meter box previously constructed.

With this method of construction, the parts are removed along with the front panel and are readily accessible for servicing or replacement. A subchassis or shelf is also often attached to the front panel or face plate and allows a considerable amount of circuit to be included. Terminal strips also can be used as anchoring points on the back side of a panel.

Purpose of Assignment

1. To construct a face plate to fit the meter box previously constructed.
2. To gain experience in planning the layout of parts and components.
3. To make an application of anodizing.
4. To gain experience in panel etching using photosensitive resists.

Tools and Equipment

Various hand tools.

Turret punch.

Squaring shear.

Greenlee chassis punch (appropriate size).

Refer to *Anodize-Dye Process.*

Refer to *Name-Plate and Front-Panel Marking (KPR).*

Materials

16- or 14-gauge aluminum, 4½ X 6½″.

Refer to *Anodize-Dye Process.*

Refer to *Name-Plate and Front-Panel Marking (KPR).*

Assignment

Construct a face plate for the meter box previously constructed and similar to the sample shown in Figure 120.

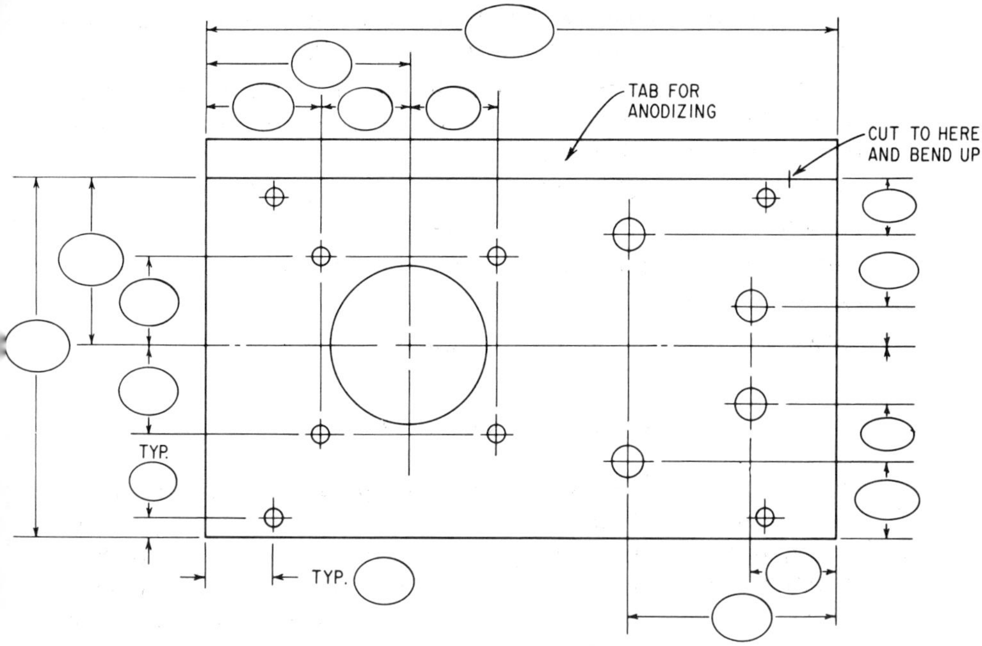

Figure 120 Sample meter-box face plate

Directions

NOTE: Omit steps 5, 6, and 7 if an anodize coating and etched labels are not required.

---

1. Determine the location and size of all holes and cutouts required for mounting components.
2. Make the layout of all centers and cut lines.

CAUTION: Are the centers of the holes for the attachment screws in the face plate going to match the locations of the swage nuts in the lips of the box? If not, the centers may need to be relocated.

---

3. Determine a *sequence* of the operations to be performed.
4. Determine the best method for making the large meter-hole cutout. The Greenlee chassis punch can buckle the face plate if

the edges are too narrow at the sides of the hole. Is there another method for cutting this meter hole? A coping saw could be used.

5. Complete all machine operations on the piece except those that should be left until after spot-etch labeling on the panel. Also leave a tab for making a connection in the anodizing solution.
6. Etch the necessary markings on the face plate [refer to *Name-Plate and Front-Panel Marking (KPR)*].
7. Anodize the face plate (refer to *Anodize-Dye Process*).
8. Complete machine operations.
9. Attach the face plate to the box and turn it in.

OPEN-TYPE CHASSIS CONSTRUCTION

A much used type of chassis is the open-type box chassis. The name does not mean that the box itself is open, but that the components (tubes, capacitors, transformers) are mounted on top of the box and not covered. This type of construction is used because it is simple, easy to construct, and is well ventilated. There are, however, certain inherent disadvantages such as the collection of dust and lack of protection against physical damage to the parts.

The type of corners involved in this construction have been reviewed in previous assignments. (refer to *Corner Sample No. 2*). Essentially, the use of a spot-weld tab and the use of lips for the attachment of the bottom cover are the main features involved in this chassis.

Figure 121 shows a pictorial view of this chassis as viewed from the bottom. Notice the spot-weld tabs and the lips for attachment of the bottom cover. The instructions for the construction of the

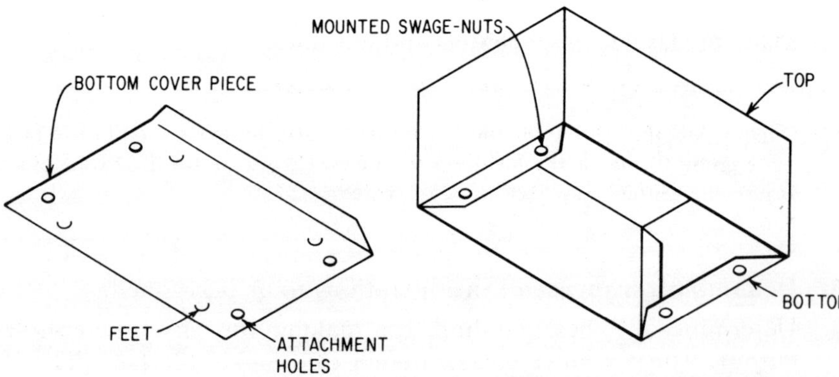

Figure 121 Open-type chassis

bottom cover (not shown in Figure 122) are given in the following assignment.

This chassis is good for power supplies, power amplifiers, oscillators, and various other circuits. In the construction of this chassis for a circuit, the specific arrangement of the parts on the chassis should be determined and included in the layout plans. The punching of holes, making transformer cutouts, and other operations should be performed on the chassis piece *before* the sides are bent up. These operations are most easily taken care of at this time.

Purpose of Assignment

1. To construct a basic type of chassis.
2. To gain experience in the use of tools in constructing a chassis.
3. To gain experience in planning and following a sequence for the operations involved in construction.

Tools and Equipment

Various hand tools.

Squaring shear.

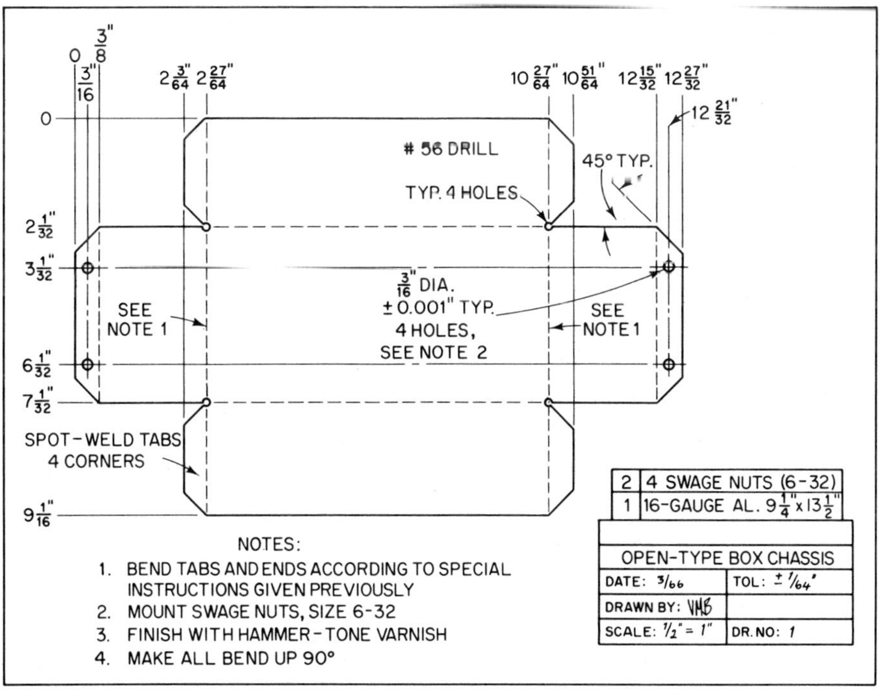

Figure 122 Box-chassis layout detail

Brake.

Turret punch.

Spot welder.

Corner notcher.

Paint spray gun and bake oven.

Materials

16-gauge aluminum sheet, 9 × 12″.

Four swage nuts, size 6-32 (Standard Pressed Steel SN632-3 or equivalent).

Hammertone varnish.

Assignment

Using 16-gauge sheet aluminum, construct an open-type chassis according to the instructions given in Figure 122.

Directions

1. If possible, decide what circuit will be built into the chassis. If possible, also determine the location and size of all the parts mounting holes and cutouts. Include these in the layout of the chassis.
2. Establish 90° reference corners on the blank stock. Make a complete layout on the stock.
3. *Outline a sequence* of the hand- and machine-tool operations to be performed. Refer to other assignments for information and work carefully.
4. Complete machine- and hand-tool operations.
5. Finish your box with an appropriate type of finish.
6. Letter-stamp your name on the box and turn it in to your instructor.

BOTTOM COVER PIECE

(continued from previous assignment)

This assignment is intended to describe a general or representative type of bottom cover piece. Variations in the shape of the piece would depend upon individual applications. Suggestions for what might be included are as follows:

1. Lips *A* and *B* (refer to Figure 123) for the purpose of strengthening the piece and holding it more rigid.
2. Relief cuts and the inset of lips *A* and *B* to avoid leaving a small gap in each corner.
3. Bolt-mounted rubber feet (mounted in hole *D*) or the type of feet depressed into the metal itself.
4. Ventilation louvers or holes.
5. Attachment screw holes (hole *C*).

Purpose of Assignment

1. To review the considerations involved in making a bottom cover piece.
2. To emphasize careful and exacting workmanship.

Tools and Equipment

Various hand tools.
Squaring shear.
Drill press or turret punch.
Brake.
Louver punch.

Materials

16-gauge sheet aluminum (appropriate size).
Rubber feet.

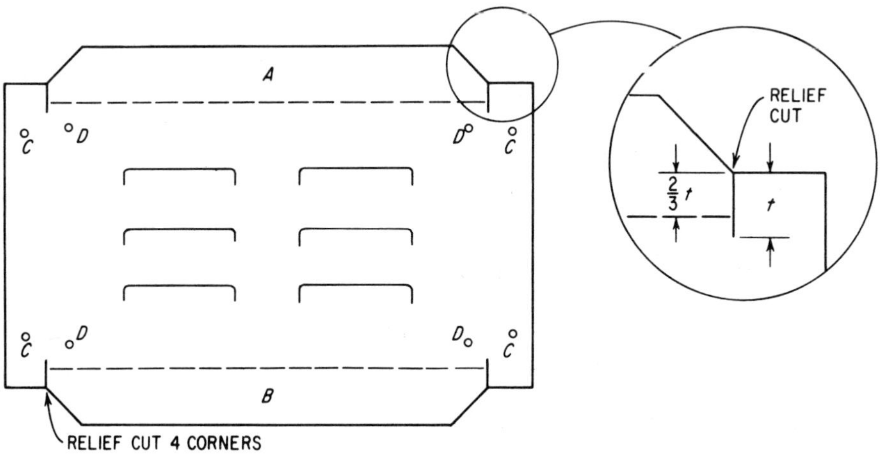

Figure 123 Bottom cover piece

Assignment

1. Make a complete layout drawing of the bottom cover piece to fit the box given in *Open-Type Chassis Construction.*
2. From this drawing, construct the bottom cover piece.

Directions

Layout Drawing

1. Determine the layout size and all dimensions for a bottom cover to fit the box given in *Open-Type Chassis Construction.* Be sure *all* necessary information is included in the drawing.
2. Consider 3/8″ lips, determine the size of all holes, and hold to a tolerance of ± 1/64″.
3. Complete the drawing in pencil on 8½ X 11″ plain paper and have the instructor check all dimensions.

Construction

1. From the information in the layout drawing made above, construct the bottom cover piece.
2. Work carefully and to a close tolerance. Complete all hand- and machine-tool operations.
3. Letter-stamp your name on the bottom cover piece, attach it to the open-type chassis, and turn it in to your instructor.

WRAPAROUND CHASSIS—TOP, SIDES, AND BOTTOM

This type of chassis is versatile, straightforward in its construction, and highly recommended as a basic type. The general structure can be broken down into two parts:

1. The wraparound frame.
2. The front and rear panels with an attached component mounting shelf.

The relationship of these parts together, forming a complete chassis, is shown in Figure 124. In this arrangement, the front panel and the component shelf are removable from the frame. This facilitates component assembly or servicing of the unit.

For the general layout of the wraparound part of the chassis, refer to Figure 125. In this layout the following items should be considered:

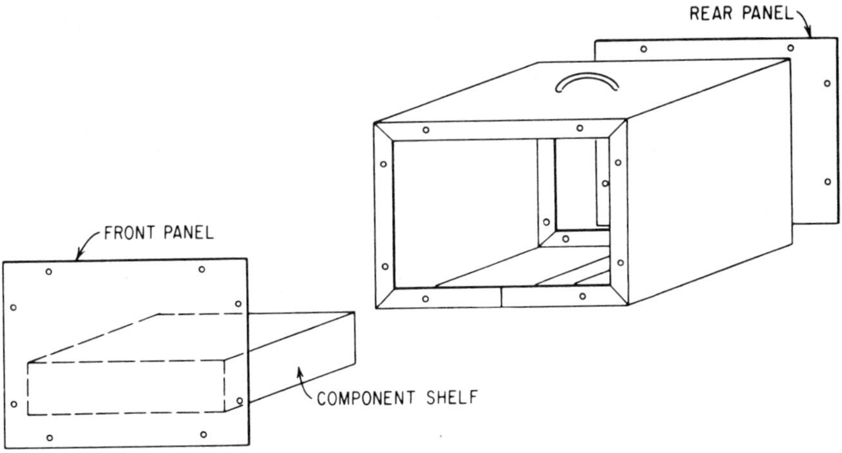

Figure 124 Wraparound chassis assembly

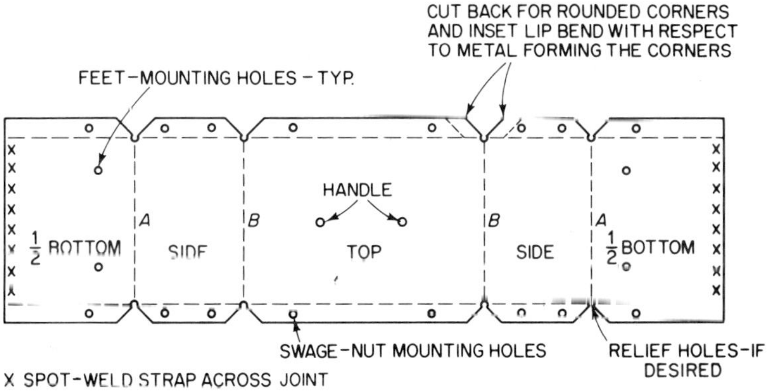

Figure 125 Layout of wraparound piece

1. The method of joining the ends of the wraparound piece in the middle of the bottom. This works out best by making the ends butt together and spot-welding a strap across the joint. This is shown in Figure 108 in the *Corner Sample No. 1* assignment.
2. The location of relief holes in each corner. These should be approximately equal in diameter to the thickness of the metal.
3. The elimination as far as possible of the gap between lips at each corner (refer to *Corner Sample No. 1*).
4. The mounting of swage nuts in the lips (refer to the information in Appendix B on *Swage-Nut Mounting*).
5. The location of holes for the feet and a handle as required.

6. Use of ventilation louvers in the sides of wraparound piece if desired.
7. The consideration of the total length of bend for the lips on each side of the wraparound piece. It is desirable to make all bends on each side in a single operation. In some cases, however, the total length of bends is greater than can be handled in the brake. In this case it is necessary to do only part of the bends at a time.
8. The proper order of the corner bends *A* and *B* (refer to Figure 125). Bends *A* should be done first and then bends *B*.
9. The method of bending the wraparound piece. It can be bent with *rounded* corners if desired. More work may be involved in order to end up with a good fit at the corners. The ends of the lips need to be cut back somewhat and the bend lines of the lips need to be inset with respect to the part of the metal forming the corners. *A round edge* on the punch in the brake is used to form a rounded corner on the workpiece.

The general layouts of the rear panel, front panel, and component shelf are shown in Figure 126. In this layout, the following items should be considered:

1. Selection of gauge of material for the front or rear panel piece. It should be slightly heavier than the gauge of material for the rest of the chassis. If aluminum is used, it should be 14- or 16-gauge, depending upon the size of the panel. In some cases, 12-gauge could be used. If sheet steel is used, it should be 16-gauge

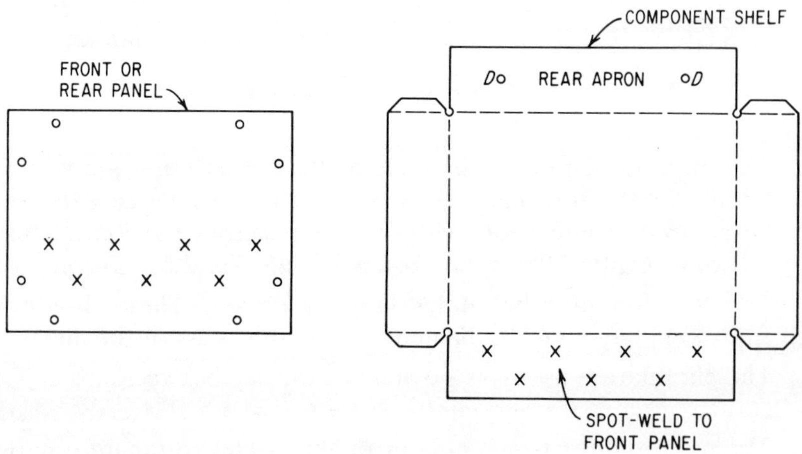

Figure 126 Front panel and shelf layout

or in some cases, 20-gauge. There should be a certain amount of rigidity, depending upon the *weight* of the components involved.

2. The determination of the exact size of the front and rear panels *after* the wraparound piece is finished. This is much easier than doing it in the reverse order.
3. The procedure for the component shelf construction as discussed in *Corner Sample No. 2*. The location of holes, etc., for mounting the circuit components should be determined before bending so that punching and cutting operations can be more easily performed.
4. The problems involved in making the front and rear panels. These are similar to those with the face plate for a meter box (refer to *Meter-Box Face Plate*). The location of the holes for controls and the mounting of components should be determined.
5. The attachment of the component shelf to the front panel. It can be done by spot welding in about six or eight spots, or in some cases by use of the control-mounting nuts.
6. The attachment of the component shelf to the rear panel for the purpose of increased rigidity. Swage nuts are mounted in holes *D* to receive a short machine screw through the rear panel.
7. The width of the component shelf. Notice that it is not quite as wide as the wraparound part of the chassis. Notches put in the attachment lips to allow the component shelf to be made wider are not satisfactory. This makes the chassis cumbersome and ill-fitting.
8. The method of marking or labeling the front or rear panels (refer to *Screen-Process Labeling* (2) or *Name-Plate and Front-Panel Marking (KPR)* (2).
9. Use of ventilation louvers in the rear panel.
10. The use of a bracket from the component shelf to the front panel if needed.

An alternative for this style of chassis can be used if desired. In some ways this might be a little better-looking and might in some ways be a little simpler in terms of the actual construction problems. Modifications other than the one shown here can also be used, with a number of very effective and interesting possibilities. Refer to Figure 127 for a view of the basic arrangement for this style of chassis.

In this arrangement the lips on the front edge of the wraparound piece are left off. This allows the wraparound piece to extend out

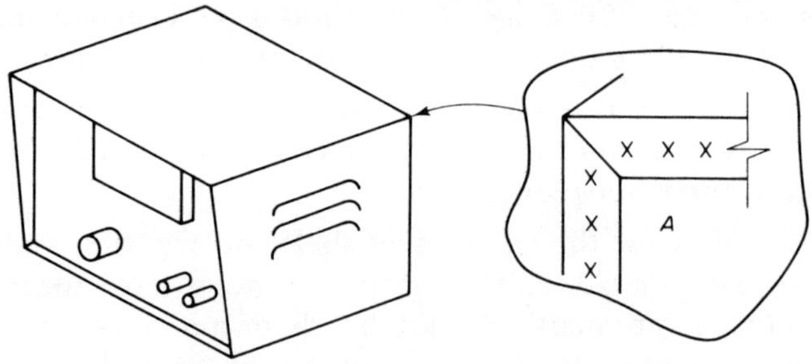

Figure 127 Wraparound chassis style

over the front panel, forming a *hood,* with the front panel being somewhat inset, giving a pleasing effect. The layout for this arrangement of the wraparound piece is given in Figure 128.

Also in this arrangement, the rear panel (*A*) can be put on the *inside* of the lips that bend down in the back, and it can be spot-welded in place to be a permanent part of the chassis box. This is shown in the inset drawing of Figure 127. Louvers or vent holes can also be punched in the rear panel, as required (see Figure 129).

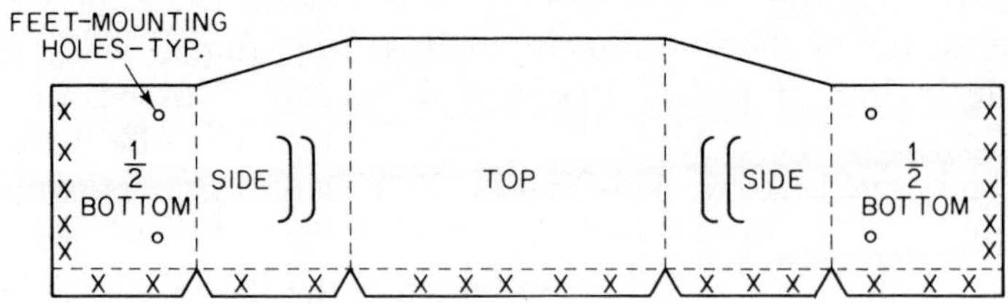

Figure 128 Layout of wraparound piece

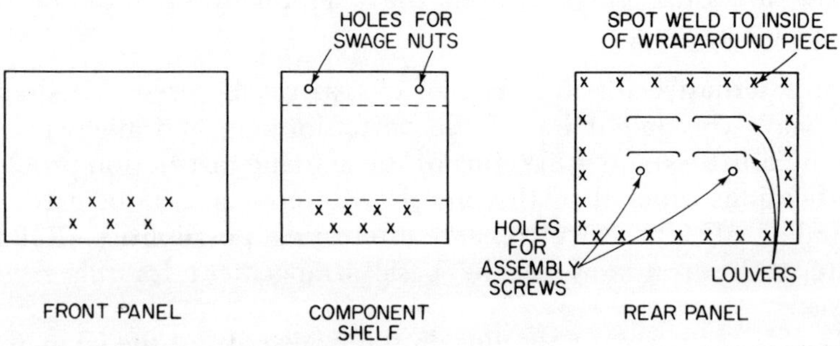

Figure 129 Front and rear panel detail

The front panel and component shelf are permanently attached together and fitted into the wraparound piece. Two screws through the rear panel to the back edge of the component shelf adequately hold the sections together. Part of the simplicity of this arrangement is in the fact that the front panel is "floating," forward and backward, and precise measurements for the depth of the component shelf do not need to be made. This is shown in Figure 130, along with other features that could be incorporated.

Purpose of Assignment

1. To show the main features of a top, sides, and bottom wraparound type of chassis.
2. To gain experience in planning the layout for one of the important basic types of chassis.

Tools and Equipment

Various hand tools.

Squaring shear.

Corner notcher.

Brake.

Turret punch.

Spot welder (see *Spot Welding*).

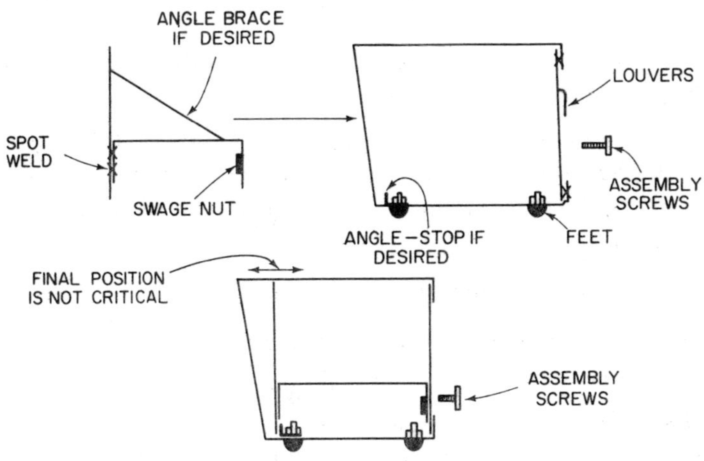

Figure 130 Side view cross section

Equipment for anodizing (see *Anodize-Dye Process*).

Equipment for panel etching (see *Name-Plate and Front-Panel Marking, KPR*).

Equipment for spray painting (see *Spray Painting*).

Materials

Sheet metal (16-gauge aluminum or 20-gauge steel, size to be determined).

Cabinet handle (optional).

Four rubber feet.

Swage nuts, number to be determined (Standard Pressed Steel SN632-3 or equivalent).

Miscellaneous machine screws.

Materials referred to in other assignments.

Assignment

1. Decide if this type of chassis can be adapted to fit the individual application. Make necessary adaptations.
2. Construct the chassis.

Directions

Preliminary Planning

1. *Think through the problems* that are inherent in this type of chassis.
2. *Determine all the layout detail,* including the location of all component mounting holes.
3. *Plan a sequence* for the completion of the hand- and machine-tool operations, using the information in the introduction.
4. *Plan a sequence* for the completion of the labeling and finishing operations.

Construction of the Chassis

1. Complete the hand- and machine-tool operations.
2. Complete the labeling and finishing.
3. Mount the parts and components.
4. Make appropriate tests and complete the unit.

WRAPAROUND CHASSIS—FRONT AND SIDES

The wraparound feature can be incorporated into a chassis having a single piece forming the front and two sides. The construction of this type of chassis is fairly simple and on the whole no more involved than any other arrangements that have been discussed. It is a recommended chassis structure.

In this type of chassis the wraparound piece for the front and sides is *permanently attached* to the component mounting shelf. These two pieces then form a rigid structure, which is the frame of the whole unit. The top and bottom pieces fit into this frame and are removable for installation of parts or for servicing. Figure 131 shows the general arrangement of this type of chassis.

For the general layout of the front and sides wraparound piece and the general layout of the component shelf, refer to Figure 132. In the construction of this type of chassis, the following items should be considered:

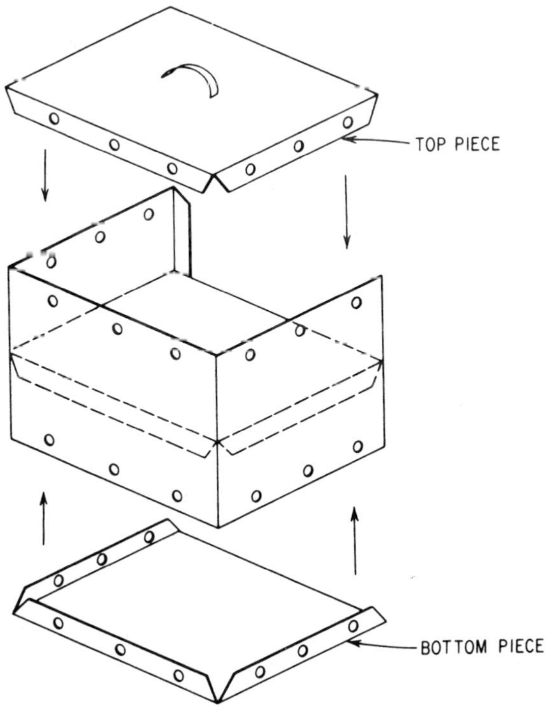

Figure 131 Wraparound chassis assembly

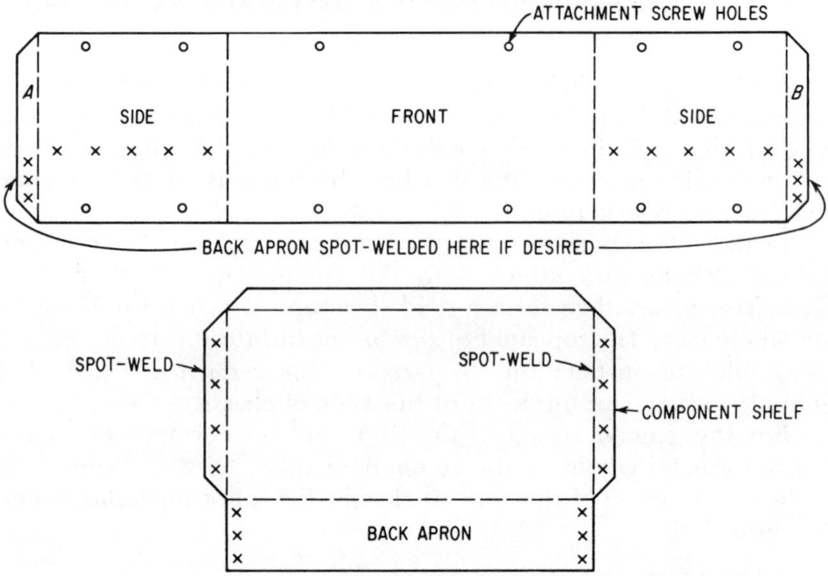

Figure 132 Layout of front, sides, and component shelf

1. The attachment of the component shelf, which fits inside the wraparound piece. This can be either spot-welded in or attached with machine screws through the sides.
2. The type of corners on the front. More work is involved when making a rounded corner. Special attention needs to be given to fitting the top and bottom pieces into rounded corners.
3. The attachment of the feet to the bottom piece and a handle to the top if desired.
4. The location and size of the holes for mounting the components in the front panel or the component shelf. This should be determined and included in the layout. Hole punching and cutout operations should be performed *before* bending and forming the chassis.
5. The modification of the layout to suit individual preferences or situations. As an example, the attachment lips can be made a part of the front and sides wraparound piece rather than part of the top and bottom pieces. The location of the heads of the attachment screws is thereby shifted.
6. The location of lips *A* and *B* on the back edge of the sides to give more rigidity. A back panel can also be attached to these lips. Do not include too many mounting screws in the plans.

7. The construction of the top and bottom pieces so they are removable for the purpose of installation or servicing (refer to Figure 133 for the general layout of the top and bottom pieces). Swage nuts are mounted in the attachment lips.
8. Determination of the kind of finish to be applied. Two colors can be used if desired.
9. The method of attaching the back panel. It can be made to slide in place or to be attached to lips on four edges (see Figure 133). Shift bend line outward if rear panel is to slide in from the side. The amount of shift will depend upon the thickness of the rear panel.
10. The use of ventilation louvers to be punched into the sides or back panel.

NOTE: Certain minute construction detail needs to be worked out on the basis of the individual application. All the detail in the layout should be determined before construction is started.

Purpose of Assignment

1. To show the main features of a front and sides wraparound type of chassis.
2. To gain experience in planning the layout for a basic type of chassis.

Tools and Equipment

Various hand tools.

Squaring shear.

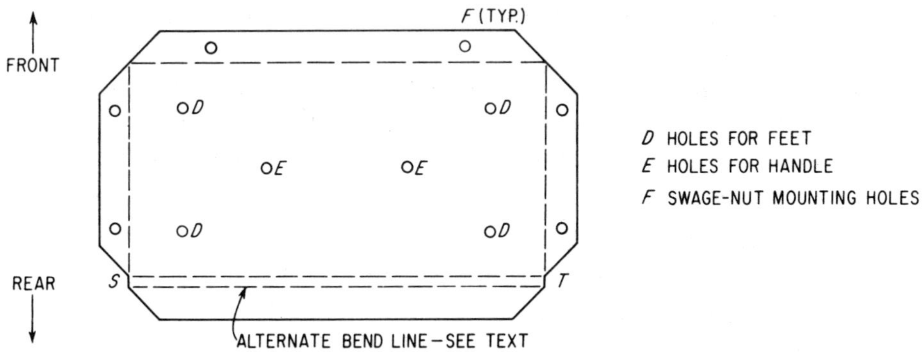

Figure 133 General layout of top and bottom pieces

Corner notcher.

Brake.

Turret punch.

Spot welder (see *Spot Welding*).

Labeling equipment to be determined.

Equipment for spray painting (see *Spray Painting*).

Materials

Sheet metal (16-gauge aluminum or approximately 20-gauge steel, size to be determined).

Cabinet handle (optional).

Four rubber feet.

Miscellaneous machine screws.

Swage nuts, number to be determined (Standard Pressed Steel SN632-3 or equivalent).

Materials referred to in other assignments.

Assignment

1. Decide if this type of chassis can be adapted to fit the individual application. Make necessary adaptations.
2. Construct the chassis.

Directions

Preliminary Planning

1. *Think through the problems* that are inherent in this type of chassis.
2. *Determine all the layout detail,* including the location of all component-mounting holes.
3. *Plan a sequence* for the completion of the hand- and machine-tool operations, using the information in the introduction.
4. *Plan a sequence* for the completion of the labeling and finishing operations.

Construction of the Chassis

1. Complete hand- and machine-tool operations.
2. Complete the labeling and finishing.
3. Mount the parts and components.
4. Make appropriate tests and complete the unit.

6

Finishes

ALODINE PROCESSING (1)

Alodine treatment of an aluminum surface is simple and easy to use. Through this process, a *corrosion-resistant* finish is formed. This finish is an ideal base in preparation for painting, although it is not a surface that will withstand wear and abrasion.

Alodine finish is formed by a chemical process. In this process, the aluminum surface is chemically converted to an insoluble aluminum compound and is therefore often referred to as a *chemical conversion coating*. This coating is desirable as indicated above, and also is a means whereby the surface is colored or tinted.

Purpose of Assignment

1. To learn the steps involved in making an alodine finish.
2. To gain experience in working carefully and accurately with a chemical process.

Tools and Equipment

Chemical tanks approximately 16 × 12 × 6″ with ventilating fan.

Water-rinse tanks.

Facilities for heating water.

Goggles for eye protection.

Materials

15 to 20% sodium hydroxide.

Bright-dip solution made up of 8 parts sulfuric acid, 4 parts nitric acid, 1 part water, and a small amount of hydrochloric acid.

Alodine crystals.

Water.

Degreasing solution.

Aluminum workpiece.

Assignment

Select the aluminum piece to be treated with the alodine coating. Carefully follow the instructions given under Directions.

Directions

Complete the mechanical operations, such as cutting, punching, forming, and bending.

CAUTION: **Use a ventilating fan over the caustic-etch tank and the acid-rinse tank. Do not breathe the toxic fumes that are given off. Be careful of the harmful solutions that are used, and wear eye protection.**

---

1. Degrease the surface, using any effective method.
2. Caustic-etch the surface. Dip in 15 to 20% solution sodium hydroxide solution for 5 to 15 minutes. Solution temperature should be from 110° to 115°F.
3. Water-rinse approximately 30 seconds.
4. Dip aluminum workpiece in bright-dip solution for approximately 20 seconds. The bright-dip solution removes the smut formed during etching and neutralizes sodium hydroxide residual. Solution is made up of 8 parts sulfuric acid, 4 parts nitric acid, 1 part water, and a small amount of hydrochloric acid.

5. Thoroughly rinse with water. Two rinse baths can be used.
6. Dip in alodine solution for 1 minute for a light coating, or 1 to 3 minutes for a heavier coating. To make the solution, mix 2-oz crystals with 1 gal of hot water. Stir well. Solution should be 60 to 70°F temperature. Avoid fumes until solution cools.
7. Water-rinse with a quick dip.
8. Dip in hot water (160 to 200°F) for not over 15 to 30 seconds. The final hot-water rinse aids in drying the piece.
9. When the finish is completed, feel the smoothness of the surface. Turn it in to your instructor for evaluation.

ANODIZE-DYE PROCESSING (2)

Anodizing is a method of electrochemically treating the surface of aluminum. It is called *anodizing* because in the process, the piece being treated is connected to the anode of a DC supply. During this process a very durable finish is formed. This finish is hard (scratch-resistant), nonporous (stain-resistant), resists corrosion, and has high electrical resistance.

The piece to be anodized is connected to the positive terminal of a DC supply and is submerged in a solution with a high *oxygen* content. A combination of the oxygen from the solution with the surface of the aluminum results in a hard thin coating of aluminum oxide. This reaction requires a current of approximately 12 A per square foot and *careful control of the solution temperature.* The result is a very hard, durable, wear-resistant finish, which can be beautifully colored with dye if desired.

Either one of two solutions is usually used for anodizing. First, *sulfuric acid* (H_2SO_4) is used with the result of a clear and somewhat transparent effect. This surface is *highly desirable* and can be used as it is, or can be dyed with a wide range of very pleasing colors. Second, *chromic acid* (CrO_3) is used leaving a slightly gray, dull appearance. This coating can be dyed also. Chromic acid may be cheaper to use, may in some cases be easier to use, and is safer to handle. This solution should be used when the piece to be anodized has folded corners, which would make it difficult to properly rinse out all of the anodizing solution. Mounted steel parts should be coated with a resist if the chassis piece is to be anodized.

Temperature control is *very important* in this process. If the temperature is allowed to rise past a certain point, the energy of the electric current is used up in electrolysis rather than in the formation of an oxide. The result is little or no anodize coating. For a sulfuric acid solution the recommended temperature is 70 ± 2°F. For chromic acid the recommended temperature is 95 ± 4°F.

The *current density* with a 15 to 18% solution of sulfuric is about 12 A per square foot. For a good coating, this means about 12 to 18 V for 20 to 30 min. Using a 5 to 10% chromic acid solution the current density will be approximately 1 to 5 A per square foot. At the proper temperature this means a potential of about 40 V for about 30 minutes.

The chromic acid anodize coating is usually not sealed as is the coating formed when using sulfuric acid. Because of this and other reasons, the sulfuric acid coating is superior and is therefore recommended. The outline for the use of the sulfuric acid solution will be given in the Directions. When using chromic acid the same steps should be followed except for the sealing of the piece.

Purpose of Assignment

1. To gain experience in the chemical treatment of aluminum pieces or chassis.
2. To learn how to work carefully and accurately in following directions.

Tools and Equipment

DC power supply.

Solution and rinsing tanks.

Rinsing tray and ventilation fan.

Water-heating facilities.

Lead plate, if the stainless-steel tank is not used.

Thermometer.

Goggles for eye protection.

Materials

Piece to be anodized.

15 to 18% sulfuric acid or 5 to 10% chromic acid.

Bright-dip solution made up of 8 parts sulfuric acid, 4 parts nitric acid, 1 part water, and a small amount of hydrochloric acid.

15 to 20% sodium hydroxide solution.

Degreasing compound or solution.

Water (demineralized).

Dye coloring, if desired.

Nickel acetate, 8 oz per gallon of demineralized water.

Sodium carbonate.

Assignment

Select a piece to be anodized. This can be the face plate constructed in *Meter-Box Face Plate.* Follow precautions and work carefully according to the instructions given in the Directions for anodizing the piece.

Directions

Be sure to allow yourself enough time to complete the job. Look ahead in the operations to be performed and be sure you will have enough tanks and equipment to complete the job. Be sure that the piece will fit into all the trays or tanks that are to be used. Put the desired scratch or polish finish on the surface *before* you anodize.

CAUTION: Work *immaculately* in all steps to prevent water spotting and unevenness of color. Do not let water spots dry on the surface and do not partially dip the piece in one of the solutions.

The sequence of operations involved in the anodize process is as follows. Omit step 3 if a clear, transparent surface is desired, and color dye is not used.

WARNING: 1. Be careful of all solutions because of possible harm to the clothes or body. Wear eye protection.

2. *Use a ventilating system* over the chemical tanks to avoid breathing toxic fumes.

3. Use plenty of water for all rinses.

---

1. Prepare the surface.

   a. Degrease and water rinse.

   b. Caustic etch with 15 to 20% sodium hydroxide, at a temperature up to 115°F, for 3 to 5 minutes.

c. Water-rinse.

d. De-smut the surface. Dip piece in bright-dip solution for approximately 20 seconds. See the materials list for the formula of the solution.

e. Water-rinse thoroughly.

2. Connect up the piece in the anodizing solution and anodize (see Figure 134). Use 15 to 18% sulfuric acid solution at a temperature of 70 ± 2°F.

a. Make the connection of the anode to the piece *out* of the solution. Leave a tab on the piece during construction or spot-weld a tab on for temporary use.

b. Connect the negative lead of the power supply to the stainless steel solution tank or a lead plate in the solution.

c. Turn on the power supply and adjust the voltage to about 5 V for 3 minutes and then increase to 12 or 18 V for 20 to 30 minutes. This will mean approximately 12 A per square foot. A good way to do this is to use three 6-V batteries in series and simply connect them to the piece to be anodized. The current will automatically regulate itself because of the resistance of the connecting wires and also the gradually increasing resistance of the piece in the solution.

d. *Circulate* the solution to prevent temperature layers from forming by using air bubbling through the solution. Also use circulating *cooling water* as needed to control the temperature level.

e. The workpiece can be hung in the solution rather than set on standoffs. The *important* and *most critical* thing is that the *electrical connection* to the workpiece be made *above* the level of the solution in the tank. The connection *cannot* be

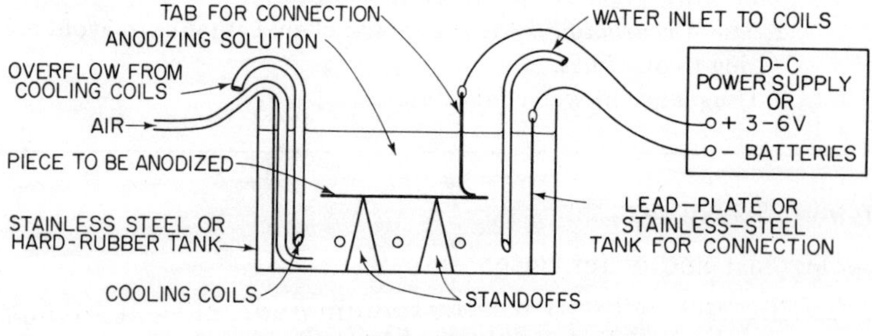

Figure 134 Arrangement of anodizing tank

made down in the solution. In most cases a 3/8″ or 1/2″ strap of aluminum can be spot-welded to the workpiece. This strap can then be used for hanging the piece and also for making the electrical connection to the piece.

3. Color-dye the piece.
    a. Neutralize the piece in a weak solution of sodium or ammonium carbonate (2 oz per gallon). Rinse with water.
    b. Dip the piece in a dye solution of 1 gram per liter of water at about 140°F. Rinse with water. The deepness of color depends upon the length of dip time, concentration of the dye, and the *depth* of the anodize coating.
    c. Seal the dye by treating the surface in a hot solution (200°F) of nickel acetate (8 oz per gallon of demineralized water) for 3 to 5 minutes.
4. Seal in hot water. Boil the piece in demineralized water for 20 to 60 minutes. This seals the pores of the oxide coating and is a *critical stage* of the process.

SPRAY PAINTING

Proper preparation of the surface is a prerequisite to good chassis painting. This preparation usually involves the complete degreasing of the surface, caustic etching, sand blasting of the surface or use of a chemical conversion coating. A prime coat is usually applied over the prepared surface and this is followed by a finish coat such as a hammertone or wrinkle varnish. Variations in the sequence and the choice of finishes can be made depending upon the individual circumstance.

Purpose of Assignment

1. To gain experience in the fundamentals of spray painting.
2. To develop habits of working carefully and neatly.

Tools and Equipment

Chemical and water-rinse tanks with ventilation fan.

Spray paint gun.

Bake oven.

Goggles for eye protection.

Materials

15 to 20% sodium hydroxide.

Degreasing solution.

Bright-dip solution made up of 8 parts sulfuric acid, 4 parts nitric acid, 1 part of water, and a small amount of hydrochloric acid.

Prime coat paint.

Finish coat paint or wrinkle varnish.

Paint thinner or solvent.

Alodine solution.

Assignment

Select a chassis or chassis cabinet to be spray-painted. This can be the chassis completed in one of the other assignments.

Directions

1. Finish all the machine- or hand-tool operations on the piece, except in some cases, the tapping of holes.
2. Use a degreasing solution to remove all grease and oils. Prior to spraying, the piece can also be wiped with a rag dipped in lacquer thinner to remove any thin grease film.
3. For all workpieces made of aluminum, prepare the surface using the following sequence:

NOTE: Omit step 3 if the piece is anything but aluminum. Do not caustic-etch if the piece contains spot-welded corners, mounted parts, rivets, or swage-mounted nuts.

---

a. Dip in 15 to 20% solution sodium hydroxide for 2 or 3 minutes.
b. Water-rinse.
c. Dip aluminum workpiece in bright-dip solution. See the materials list for the formula of the solution.
d. Water-rinse.
e. Dip the piece in an alodine solution if available, 1 to 2 minutes.
f. Rinse with clean water.
g. Dip the piece in hot water 15 to 30 seconds.

4. Determine a method of holding or hanging the piece while spraying and baking. A small bracket may need to be made.
5. Mask off areas where paint is not desired. Mix paint thoroughly. Spray on the prime coat and bake at 250°F for 20 minutes.
6. Spray the finish coat on the piece evenly and smoothly, watching for runs and thin spots. Spray the hard-to-get-at areas first and then the open areas. Use a wrinkle varnish or a hammertone finish (check the instructions on the can for additional information). Bake at 225°F for 20 minutes.
7. Clean up. Clean the top rim of the paint can before closing. Thoroughly clean up the spray gun and any brushes that might have been used.

7

Labeling and Marking

SCREEN-PROCESS LABELING (2)

The photographic processes can be used in conjunction with screen-process techniques to achieve a fine printed pattern detail. By using this combination, front panels can be professionally labeled, printed-circuit etching resist patterns can be achieved, and components in a chassis or on a printed-circuit board can be neatly identified.

In recent years new films have been developed for use in the screen-process method of printing. These films are simpler to use, require less chemicals and equipment, are easier to handle, and give very excellent results. The process has been simplified to the point where all that is required is that the film be exposed through a positive of the pattern and developed in cold running water before it is placed onto the screen. In the past, most of the work included preparation of the film, and careful handling of the film in getting it placed onto the screen. Now it is a much more straightforward process than it used to be.

The process basically includes exposing the film through a

pattern, developing the film with cold water, making up the screen stencil, and then printing the pattern. The total process involves considerable time because of drying requirements, but it is a method that gives good results.

The screen-process printing method is often used to *print* the resist pattern on a printed-circuit board. In industry this method is quite satisfactory on the mass-production line where many copies of the same pattern need to be produced. When used in conjunction with photographic processes, consistently accurate patterns with sharp detail can be obtained.

On the mass-production line a large screen stencil is made up with the pattern repeated several times on the screen area. Large sizes of printed-circuit board are then stenciled and at a later stage this board is cut apart to form the individual circuit-board pieces. The overall sequence is basically the same when setting up to run a single copy.

It should be especially noted here that a photographic film *positive* is used for exposure. This means that the conductor pattern on the film will show up as a solid wide line and there will be a clear transparent background. The background areas then will be exposed and the conductor areas will be unexposed. In development of the film the conductor areas will dissolve, leaving them clear and free to allow the paint or ink to seep through during the actual printing. This paint or ink on the conductor areas will act as a resist in the etchant (see Figure 135).

The same sort of thing occurs when printing a front panel or

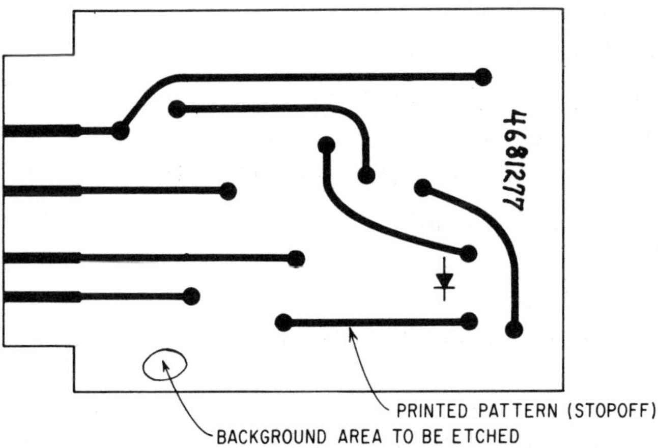

Figure 135 Screen-process printed pattern

labeling a circuit board. On a front panel the paint will seep through where the labeling is.

Purpose of Assignment

1. To show one method of marking, labeling, or applying an etchant resist.
2. To become acquainted with photographic screen processes.

Tools and Equipment

Screen-process frame with approximately 150- to 200-mesh nylon fabric, and base with clamp hinges.

Printer's squeegee.

Semidark room with ultraviolet filtered light.

Contact printing frame with ultraviolet light.

16- or 18-gauge aluminum piece slightly larger than the size of the film.

Circulating fan.

Gloves for use when cleaning up the paint.

Spatula for handling paint.

Materials

Film to make a positive of the pattern.

Panel to label.

Screen-process film, autotype superfilm.

Cold running water.

Water-soluble screen filler.

Lacquer thinner or butyl Cellosolve.

Screen-process paint, epoxy paint, or plating resist.

Brown tape, water-base adhesive on the tape,

Wiping rags and scrub brush.

Masking tape.

White drawing paper.

Typeset letters or printed letters with sharp clear outlines.

Assignment

Label a front panel for an instrument under construction.

Directions

Cleanliness is an important part of successful photographic work. Remember to work carefully and keep materials and equipment clean.

1. Determine the pattern to be printed. Develop the artwork to an exact double size using black on a white background. Use drawing instruments to make lines, circles, or other patterns. The letters and numbers can be typeset and then cut out and glued in place. Any printed letters with sharp and clear outlines can be used (see Figures 145 and 146).
2. Make a film positive, reduced 50%. If necessary, refer to a photographic techniques manual or have the film made commercially.

NOTE: If photographic equipment is not available, the film can be made commercially, reduced to full size. Also, if the above artwork is done at a 1:1 size on a transparent plastic film, a Color Key film can be used for making the positive. The pattern will need to be made on the Color Key film twice. The first time it comes out negative and the second time it comes back to a positive. The original artwork probably would not be quite right to use as a positive for exposing the screen-process film except in some cases when press-on letters or numbers are used, and a clean positive original is made.

---

3. Place the following items in order in the contact printing frame (see Figure 136).

   a. Screen-process film. See the materials list for the type of film to use.

   b. Film positive.

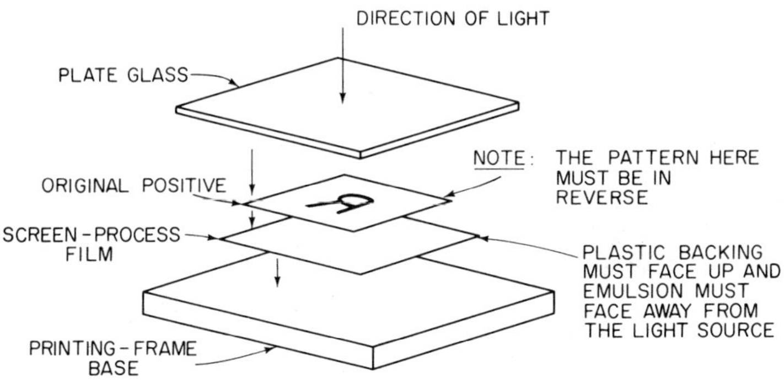

Figure 136 Preparation for exposure of film

c. Plate glass for pressing the film positive tight against the screen-process film. Handle the film in a room with ultraviolet filtered light.

NOTE: The screen-process film *must* be facing in the correct direction with the plastic backing facing the light source and the emulsion *must* face away from the light source. Also note that the film positive *must* be in reverse here because of the way the stencil is placed on the screen. At this point it is reversed again, making it come out in the correct direction for printing.

---

4. Expose the pattern for 3 minutes using a proper ultraviolet light source. This can be direct sunlight, a carbon-arc light, or a No. 2 photoflood bulb for 4 minutes with the bulb at 30″.
5. Develop the film by letting cold water run gently across the emulsion of the film until the unexposed areas of the pattern are completely washed clean. It is fairly easy to tell when this occurs if the film is placed, emulsion side up, on a smooth shiny piece of sheet aluminum (refer to Figure 137). Look for the effects of over- or underexposure.
6. Attach the film to the screen. Place the screen upside down and position the film in the center of the screen with the emulsion facing the screen (refer to Figure 138).
7. Turn the screen right side up on a soft pad and blot across the screen, gently pressing the emulsion into the fibers of the screen (see Figure 139). *Do not* rub across the pattern, which would cause the emulsion to smear. Put a sheet of paper over the pattern and gently blot the water from the screen while pressing the emulsion into the screen fibers.

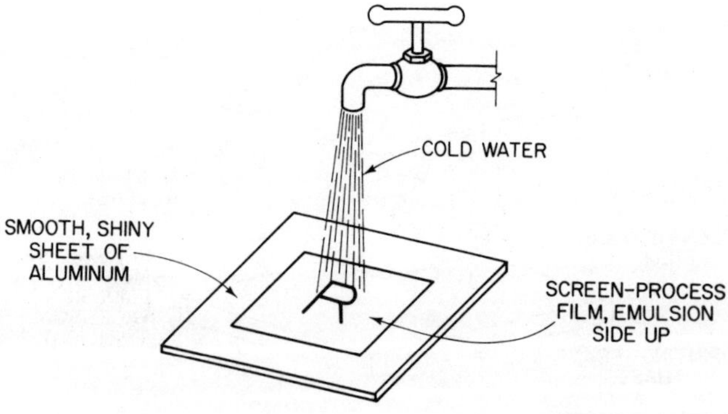

Figure 137 Developing screen-process film

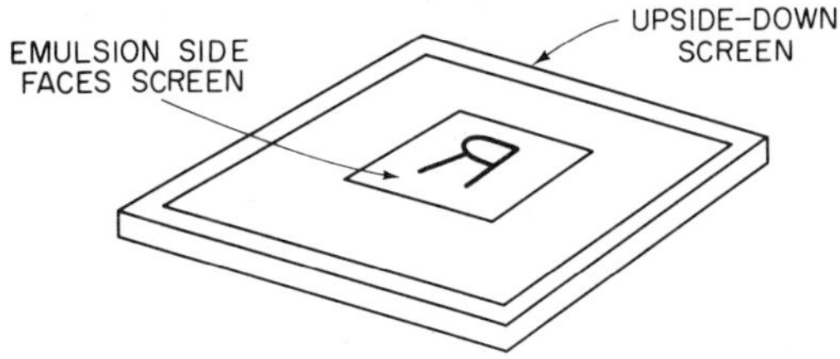

Figure 138 Stencil film placed on screen

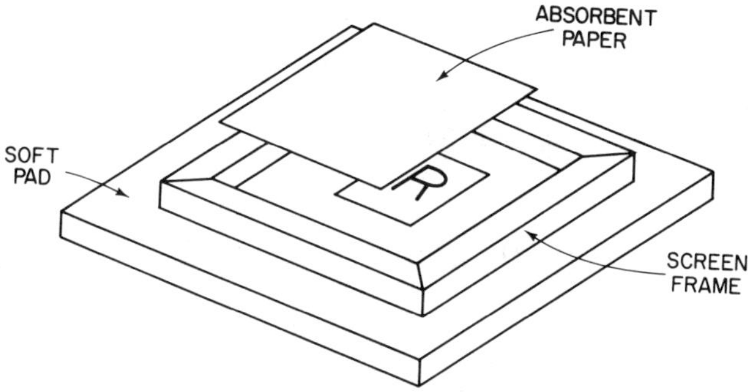

Figure 139 Preparation for pressing emulsion into fabric

8. The film and screen now are allowed to dry. A fan can be used to speed up the drying process or it can just sit overnight.
9. *After* the film is dry gently peel off the plastic backing from the film. The emulsion should stay embedded in the fibers of the screen, making a stencil.
10. Using some brown paper tape, put a protective covering over the wood portions of the frame and onto the edge of the fibers of the screen (see Figure 140). Cut the tape into strips to fit the distance between edges. Eight strips of tape should be enough.
11. With the screen facing up, mask out the areas not to be printed. This mostly includes the background area surrounding the film pattern. Use a water-soluble screen filler and apply with a pushing action as shown in Figure 141. The pusher is a cardboard or plastic piece with rounded corners to keep from harming the screen fibers. Let the filler dry.
12. Place the screen frame in the hinge clamps on the printing-frame base. Set up registry tabs to position the panel or board under the screen (refer to Figure 142). Prepare the panel or board surface to be clean and free of grease film.

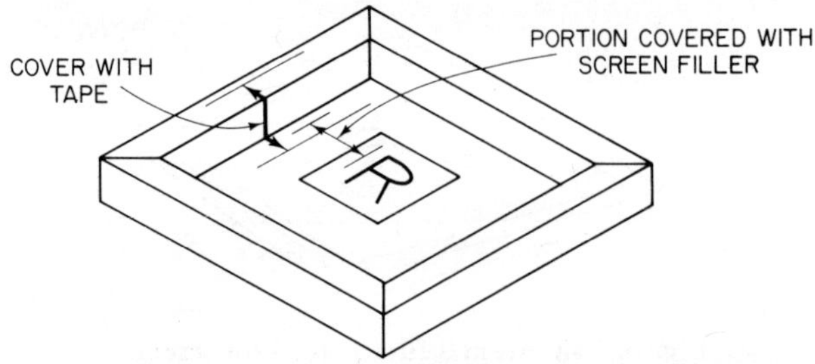

Figure 140 Protective coverings on frame and screen

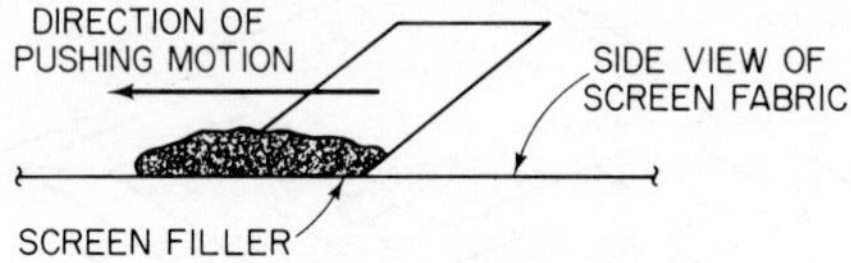

Figure 141 Applying the screen filler

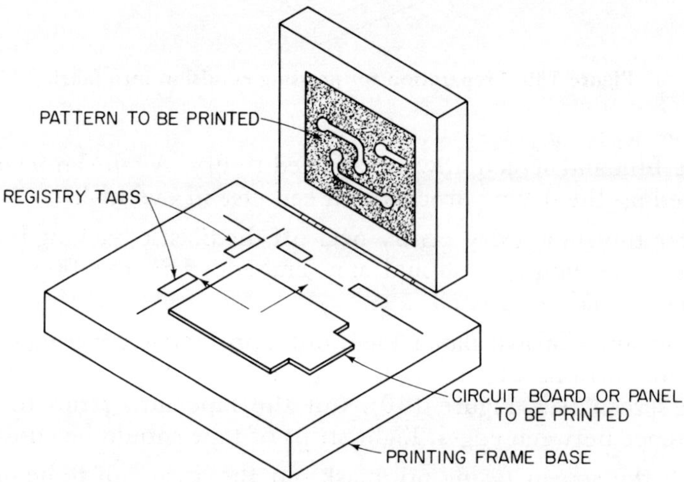

Figure 142 Registry tabs used to position board

13. Run printing tests on scrap paper. The paint consistency should be about like heavy cream. Sweep the squeegee once across the screen for each print, carefully controlling the pressure (refer to Figure 143). Use an *epoxy-base* paint if it is a front panel that is to be printed or a *plating resist* if it is a printed-circuit board that is to be etched.

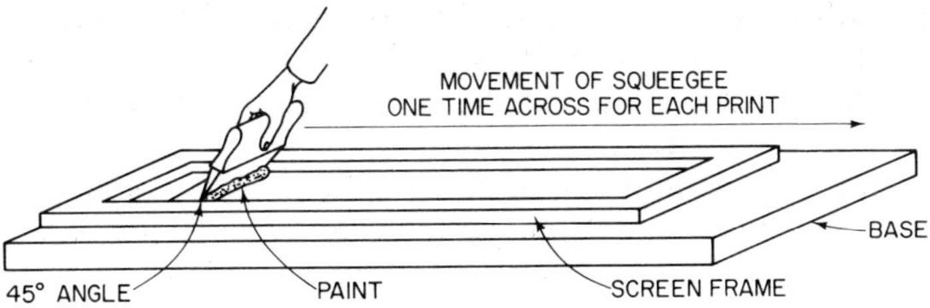

Figure 143 Squeegee movement across screen

14. Print the final pattern if the printing tests are successful.
15. When finished, clean the paint out of the screen. Use *butyl Cellosolve* if the epoxy-base paint was used, or *lacquer thinner* if the plating resist was used. Put the screen on a clean smooth metal surface and pour a small amount of the cleaning solution on top of the screen. Use rags to wash the paint off the screen and to soak up the cleaning solution. Use adequate ventilation. Also clean the squeegee and other equipment.
16. If the pattern is not to be used again, wash the screen filler, the brown tape, and the stencil emulsion out of the screen using a brush and hot water. The screen can be soaked for a while if desired. Brush and rinse the screen to get all traces of material out of the fibers of the screen. Dry the screen.
17. Oven-bake the panel or board to dry it, using a low-temperature oven.

NAME-PLATE AND FRONT-PANEL MARKING (KPR) (2)

Lines, numbers, and letters can be marked on front panels in a number of ways, including stamping, decals, engraving, silk screen, etching, and by the use of a photosensitized surface. Each of these ways has *advantages* and *disadvantages,* which need to be weighed to determine the method best suited to a particular set of circumstances. The factors to be considered are as follows:

1. Cost.
2. Time required.
3. Required sharpness of detail.
4. Equipment available.
5. Number or complication of markings.
6. Materials or finishes involved.

An outline will be given in this assignment for marking a front panel by the use of *photo-resist materials* and a *panel etchant*. By this means a detailed or complicated pattern can be effectively inscribed on a panel or name plate.

Because the equipment in most school shops is limited, it is recommended that on a large panel, only a small area be etched at a time. Use a *spot technique* and etch only a small group of numbers, lines, or letters. Further, it is suggested that the etchant resist be applied only to the side where the etching will take place and that an *open reservoir* for the etchant be devised. This can be an open-ended container or a wall of clay built around the spot where the etching will take place. A low-walled reservoir can also be built by using thin strips of masking tape. This is shown in Figure 144.

Purpose of Assignment

1. To introduce a good method of marking a front panel or a name plate.
2. To learn about the characteristics of photo-resist materials.
3. To gain experience in the techniques of handling and using photosensitive resist materials.

Tools and Equipment

Chemical tanks approximately 16 × 12 × 6″ with ventilation fan.

Water-rinse tanks.

Bake oven.

Exposure frame.

Goggles for eye protection.

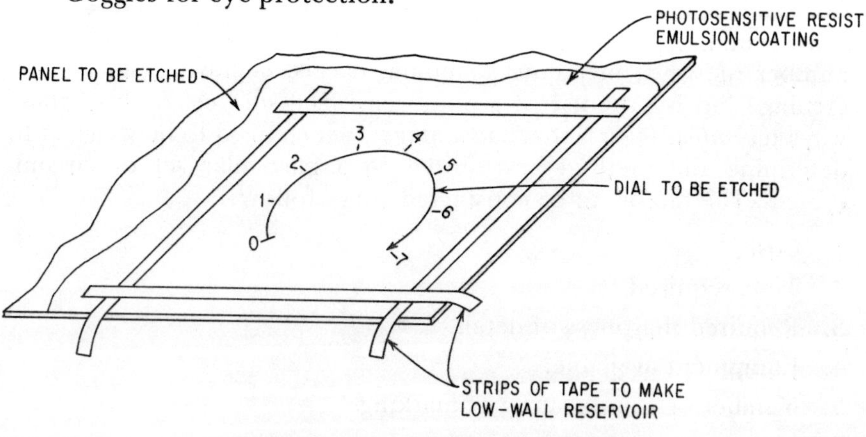

Figure 144 Reservoir for spot etching

Materials

Photographic film positive.

KPR (Kodak Photo Resist).

KPR developer.

15 to 20% sodium hydroxide and ferric chloride etchant.

Bright-dip solution, made up of 8 parts sulfuric acid, 4 parts nitric acid, 1 part water, and a small amount of hydrochloric acid.

Degreasing solution.

Filler paint.

Garnet paper—400 grit.

Name plate or front panel to be etched (usually made of aluminum).

Masking tape.

Assignment

Etch the labeling on a name plate or front panel. This can be the face plate constructed in the *Meter-Box Face Plate* Assignment. Follow the sequence of steps necessary for the etching process which are given in the Directions.

CAUTION: Be extremely careful of the chemicals. Wear an apron, rubber gloves, and use a pair of tongs to handle the piece. Also, be sure that the ventilation fan is turned on so that you will not breathe toxic fumes.

Directions

1. Determine the pattern to be etched and develop the artwork. The pattern should be double-sized, black on white background. Use drawing instruments for lines, circles, or other patterns. Typeset the letters and numbers. Cut out and glue them in place as shown in Figure 145. Any printed letters with sharp clear outlines can be used.
2. Make a film positive reduced 50%. (See Figure 146.) If the equipment is not available, have a local firm make the film positive.
3. Prepare the surface of the panel to be etched.
   a. Degrease, using any effective method and water rinse.
   b. Caustic-etch the surface with 15 to 20% sodium hydroxide solution at a temperature up to 115°F. Etch for approxi-

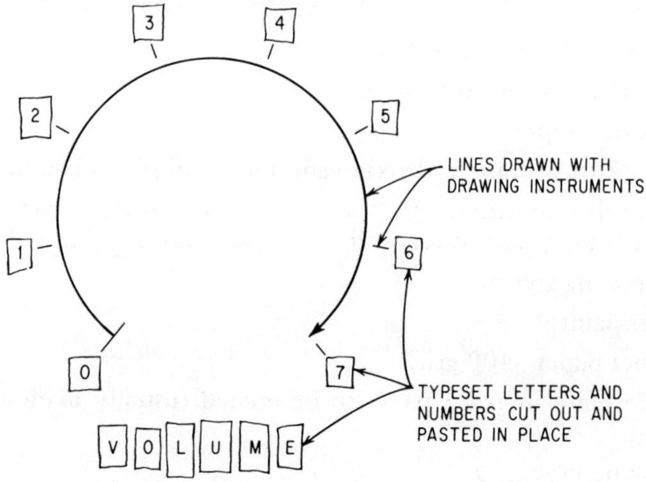

Figure 145 Artwork photomaster

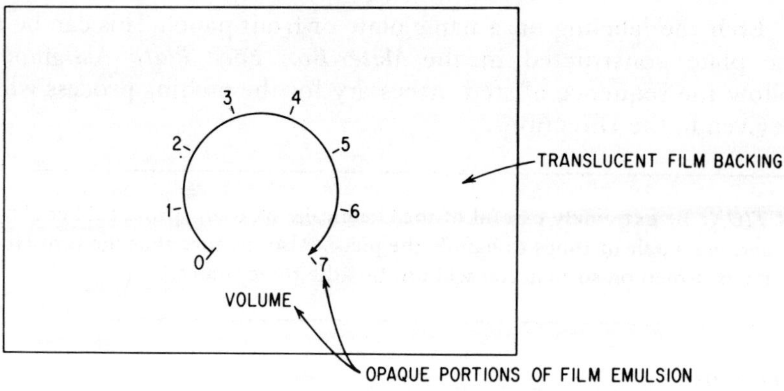

Figure 146 Film positive used for etching

mately 3 to 5 minutes, depending upon the temperature and concentration of the solution.

c. Water-rinse.

d. Bright-dip the panel for about 20 seconds. The formula for the solution is given in the materials list. This step removes smut and neutralizes the sodium hydroxide residue.

e. Water-rinse thoroughly and dry.

f. Apply a conversion coating if desired. The instructions for this step would come from the manufacturer of the materials used for this. It is not always a necessary step.

g. Water-rinse and dry.

NOTE: In some cases steps 3a to g can be bypassed by thoroughly sanding the surface with a fine grit (400 to 600 grit) garnet paper. Use waterproof paper and flush with water while sanding. Dry thoroughly.

---

4. Apply the KPR coating. Use any effective method such as brushing, spraying, dipping, or use a roller. Air-dry the coating for 2 minutes and then prebake for 10 minutes at 140 to 200°F.
5. Expose the pattern. Use a proper light source such as a carbon arc light, direct sunlight, or an ultraviolet light in a contact printing frame. Exposure time would be about 4 minutes in direct sunlight. Determine the proper amount of time by experimentation. Try long and short exposures on prepared test pieces before exposing the workpiece.
6. Develop the image.
   a. Develop for 3 minutes in KPR developer, with slight agitation.
   b. Take out of developer and quickly rinse gently under cold running water. Rinse carefully and do not touch the soft emulsion.
   c. Look for effects of underexposure or overexposure. The *background is spotty* if underexposed. The *line is lost or narrowed* if overexposed.
   d. Postbake for 10 minutes at a temperature of 300 to 450°F.
   e. Check the background areas carefully. If necessary, recoat part of the background area to ensure complete coverage.
7. Etch the panel. Use the proper solution for etching. For aluminum, use ferric chloride (40° Baumé) (this is the recommended solution), sodium hydroxide (20%) at a temperature up to 115°F, or a 1:4 solution of hydrochloric acid at a temperature up to 150°F (cooling may be required).
   a. Use a spot technique and do not try to do the whole panel at once. Build a small reservoir around a small area and complete the etching in that spot. This is illustrated in Figure 144.
   b. Stop the process shortly after it starts, to check for flaws in the pattern or background area.
   c. Etch for the length of time that will produce a pattern at least 0.005″ deep.
   d. Water-rinse and dry.

8. Repeat the process for the other sections of the pattern until all sections are completed.
9. Apply the filler if desired. Use black paint or regular filler material to make the letters stand out. Dry-bake for 15 minutes at a temperature of 200°F. If the panel is to be anodized, apply the filler after anodizing.
10. Remove the resist and filler coating. Use any method such as scrubbing with developer, using a stripper, a fine garnet paper, or steel wool. There are several problems in removing the resist coating and excess filler. These need to be worked out on an individual basis. If a scratched surface is acceptable, the use of a waterproof garnet paper (400 to 600 grit) and water works very well. In some cases, the use of ferric chloride as an etchant will cause the etched pattern to be very dark in color, eliminating the need for the use of a filler.

PART FOUR

PROJECTS

8

Construction and Assembly of Projects

PRINTED-CIRCUIT TEST-BOARD FABRICATION (1)

This printed-circuit-board assignment has to do with the following operations:

1. The fabrication of a sample printed-circuit board.
2. Mounting turret terminals.
3. Soldering jumpers and components to the printed-circuit pattern.

It is assumed here that certain instruction has been received by the students, other assignments have been completed, and that this assignment could be a "midterm test" that is very strictly graded. The grading of this board can be done on the basis of the following items:

1. Size of board.
2. Cleanness of board.
3. Board edges.

4. Conductor pattern etching.
   a. Undercutting.
   b. Pattern background.
   c. Pattern imperfections.
5. Tin coating on conductors.
6. Soldering.
   a. Pads covered—holes in solder nonexistent.
   b. Amount of solder.
   c. Cold solder connections.
   d. Properly clipped leads.
7. Component orientation.
   a. Color, number, or word identification orientation.
   b. Component-lead bending.
   c. Mounting of components straight, parallel, or square with board edge and centered between bends.
   d. Jumpers straight and close to board.

Figure 147 shows a typical *printed-circuit-board component mounting schedule.* Information relating to the location of each component would come from a table similar to this one. Figure 148 shows a typical *printed-circuit-board drill schedule.* The size of the drills should be 0.008 to 0.012″ larger in diameter than the diameter of the component lead that is to feed through the holes. The turret terminals should use a hole size only about 0.002″ larger than the shank of the terminal.

Purpose of Assignment

1. To gain experience with the complete fabrication of a printed-circuit board.
2. To gain experience in soldering to a printed-circuit board.

Tools and Equipment

Various hand tools for soldering.

Soldering vise.

Soldering device with dampened sponge.

Air hose.

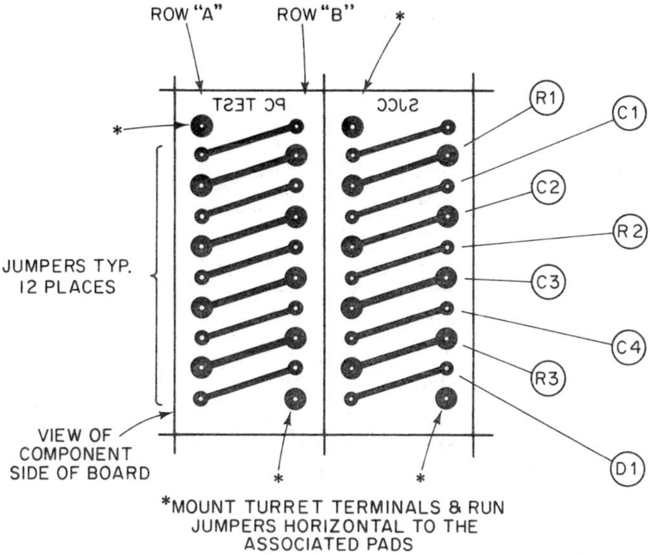

COMPONENT MOUNTING SCHEDULE	
R1	
R2	
R3	
C1	
C2	
C3	
C4	
D1	

Figure 147 PC test board component mounting schedule

Equipment for printed-circuit-board etching and fabrication.

Turret-terminal mounting block and punch.

Materials

Printed-circuit test-board negative. (If a negative is to be made for the circuit pattern of this assignment refer to direction steps 5 and 6 of the *Spiral-Sweep-Generator Artwork* (1) assignment or the *Printed-Circuit Artwork Development* assignment.)

Single-sided printed-circuit board, 3 × 3½″.

20-gauge jumper wire.

4-turret terminals.

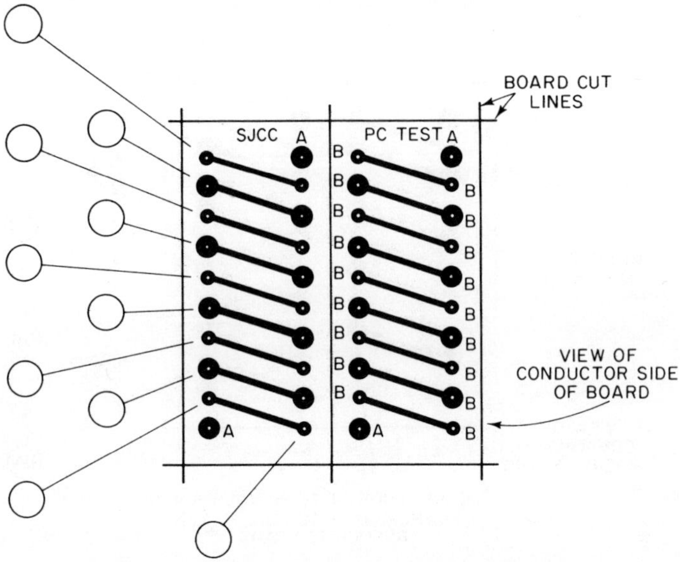

DRILL SCHEDULE		USE OF HOLE
A		Turret Terminal Shank
B	#60	# 20 Ga Jumper Wire
C		Small Lead Components
D		
E		

Figure 148 PC test board drill schedule

Miscellaneous components as assigned by the instructor.

Liquid flux and isopropyl alcohol, flux remover, or Freon vapor degreaser.

22-gauge rosin-core solder.

Assignment

Fabricate the printed-circuit test board and mount turret terminals, jumper wires, and components according to information given in the directions.

Directions

1. Clean the board, etch, and tin-coat both sections of the pattern of the printed-circuit test board. Refer to other assignments for information on these operations.

2. Fill out the component mounting schedule (Figure 147) according to your instructor's directions using the assigned components.
3. Determine the proper drills to use (refer to the introduction and the *Mounting Printed-Circuit Components* assignment). Fill out the drill schedule and the blank spaces on the test-board pattern in Figure 148 according to your instructor's directions, using the proper drills and letters to match (see the introduction).
4. Drill the board, referring to the drill schedule (Figure 148).
5. Trim the board on the inside edge of the cut lines. Leave both sections of the board together and file the edges of the board to be perfectly smooth.
6. Mount the turret terminals (refer to *Printed-Circuit-Board Soldering*).
7. Install the jumpers first. Use both methods of soldering leads to a printed-circuit board. (These methods are described in the introduction of *Printed-Circuit-Board Soldering*).
    a. Use the first method for soldering to the pads in row A, Figure 147.
    b. Use the second method for soldering to the pads in row B, Figure 147.
8. Mount the components using either method of making the lead connections to the board (refer to Figure 147). Be sure to orient the components so that value markings are readable and color bands facing in the same direction. Use a lead bender if available (refer to Figure 149).
9. Clean the board thoroughly using alcohol, flux remover, and the air hose or wiping tissue, or Freon vapor degreaser.
10. Put your name on the board with tape and turn it in to your instructor.

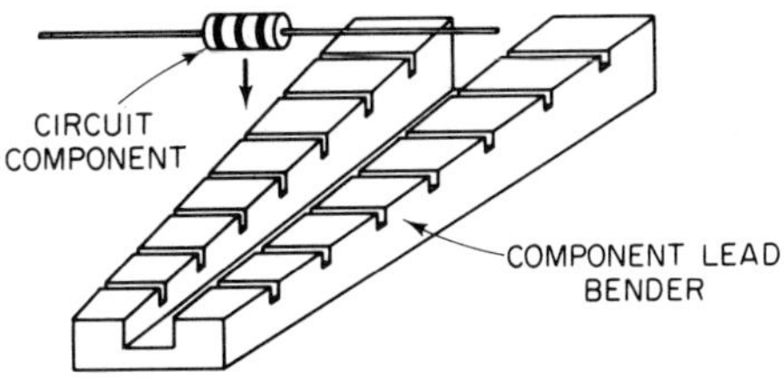

Figure 149 Use of component lead bender

THE BINARY HALF-ADDER (1)

With the rapid advancement of computer technology today, it is important to continually look for new information relating to the state of the art in electronics. This assignment will include an introduction to basic computer theory as well as the assembly of integrated circuits and light-emitting diodes. Figure 150 shows a view of some commonly used integrated circuits and a view of a typical *light-emitting diode* (LED). The diodes come in yellow, green, and red.

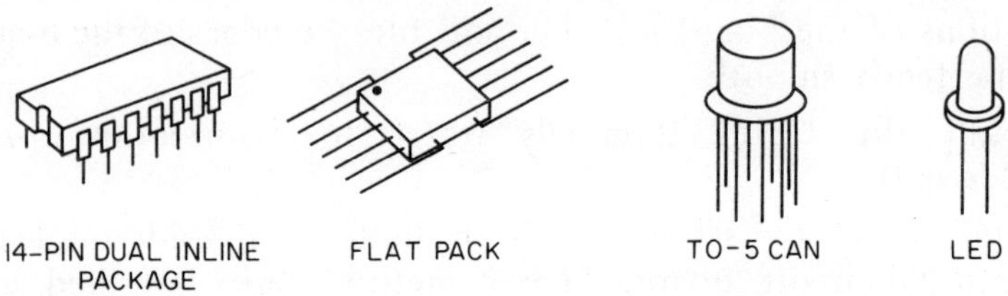

Figure 150 Integrated circuits and LED

In computers many types of relatively simple circuits are put together to build up the complete system. The NOR gate, NAND gate, and amplifiers are examples of basic circuits which are the building blocks of more elaborate system circuitry. These types of circuit are built into small chips referred to as *integrated circuits* (ICs). Often these chips include dozens or hundreds of transistors, resistors, diodes, and capacitors. Figure 151 shows the circuit symbol of the NOR gate and NAND gate and the basic circuit symbol for the 962 IC chip used in this assignment. The actual circuitry included in the 962 flat-pack package is shown on the right.

The operation of most counting or computing circuitry uses binary numbers. For entry or retrieval of information in or out of a digital system, octal numbers are used.

A binary number system is a code system that uses only two symbols. These symbols can be anything like "yes" or "no," "+" or "-," "true" or "false," or just simply "1" or "0." Because there are only two symbols in the binary number system, we say it is operated on the base of 2. With this in mind we can represent numbers in binary (8–4–2–1 code) as follows:

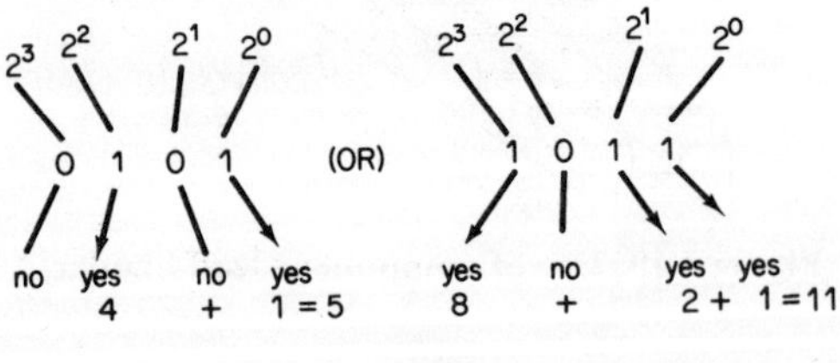

Also, with this in mind, we can count as follows:

0, 1, 10, 11, 100, 101, 110, ...
↓ ↓ ↓ ↓ ↓ ↓ ↓
0, 1, 2, 3, 4, 5, 6, ...

Four simple cases describe binary addition. This is not ordinary arithmetic, even though it may seem as though it is. It is binary addition, and goes as follows:

a.	b.	c.	d.
0	0	1	1
+ 0	+ 1	+ 0	+ 1
0 ← sum	1 ← sum	1 ← sum	10 ← sum, ← carry

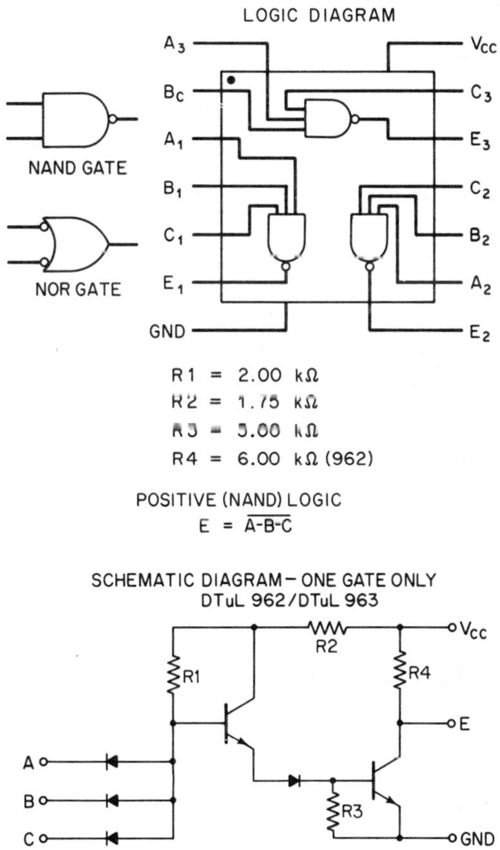

Figure 151 Symbols and circuit of the 962 Dip

A four-place term with four bits is used to represent each decimal digit. Using four binary bits, 16 possible combinations of "0's" and "1's" can be found ($2^4 = 16$). But since there are only 10 digits (0 to 9) in the decimal system, six of the combinations turn out to be "invalid." This means that when the sum of two BCD digits is greater than 9, the correction of adding 6 (0110) must be done. A full-adder circuit is required for this because of three inputs being involved.

Example: 8 + 5 = 13

```
          1000
         +0101
          1100 . . . . . . . . invalid because 13 is two digits
         +0110 . . . . . . . . correction (add 6)
 0001     0011

    1        3   = 13
```

The binary half-adder is a logic circuit with two inputs and two outputs. The circuit adds two bits at a time, yielding a "sum" and a "carry" output. This half-adder circuit is built using LEDs as the readout, indicating the "sum" and "carry."

The construction of this circuit includes mounting the LEDs and ICs and fabrication of the battery holder and board. It is suggested that this project can be a "final exam" test, graded strictly on the following points:

1. Size of board.
2. Board edges.
3. Cleanness of board.
4. Pattern etching.
   - a. Undercutting.
   - b. Pattern edges and detail.
   - c. Tin coating.
   - d. Background.
5. Parts mounting.
   - a. Straightness of all parts.
   - b. Orientation of component values.
   - c. Switches.

6. Soldering.
   a. Appearance.
   b. Amount.
   c. Over-heating of the board.
   d. Other.
7. Battery holder.
   a. Deburred edges.
   b. Bends and outline.
   c. Alodine finish.
   d. Mounting of holder.
8. Operation.

Purpose of Assignment

To become acquainted with elementary digital circuitry and the assembly of circuits, including ICs and LEDs.

Tools and Equipment

Various hand tools for soldering.

Soldering device with dampened sponge.

Soldering vise.

Air hose.

Equipment for printed-circuit board etching and fabrication.

Safety glasses.

Materials

Single-sided printed-circuit board, 2¼ X 4½″.

2–LEDs.

2–ICs #962.

2–220-Ω resistors, 1/4 W.

Battery connector.

18-gauge aluminum, 2¼ X 3½″.

Material for homemade switches (spring brass).

2–4-40 X 1/4″ bolts, nuts, and lock washers.

9-V battery.

Liquid flux and isopropyl alcohol, flux remover, or Freon vapor degreaser.

22-gauge rosin-core solder.

Assignment

Fabricate the binary half-adder printed-circuit board and mount the components for completion of the circuit. Refer to the Directions. (*Note:* All components except battery will be mounted on the conductor side of the board.)

Directions

1. Obtain the binary half-adder circuit negative. (See Figure 157. If a negative is to be made for the circuit pattern of this assignment, refer to direction steps 5 and 6 of the *Spiral-Sweep-Generator Artwork* (1) assignment or the *Printed-Circuit Artwork Development* assignment.)
2. Clean the board, etch, tin coat, drill, and trim the printed-circuit board. Refer to other assignments for information on these operations (refer to the drill schedule, Figure 158).
3. Mount the IC flat packs. Note the following steps:
   a. Prepare the flat packs for mounting. Put the lead of a 1/2-W resistor between the flat pack and a flat surface and press the flat-pack leads down against the flat surface using a square edge (see Figure 152).

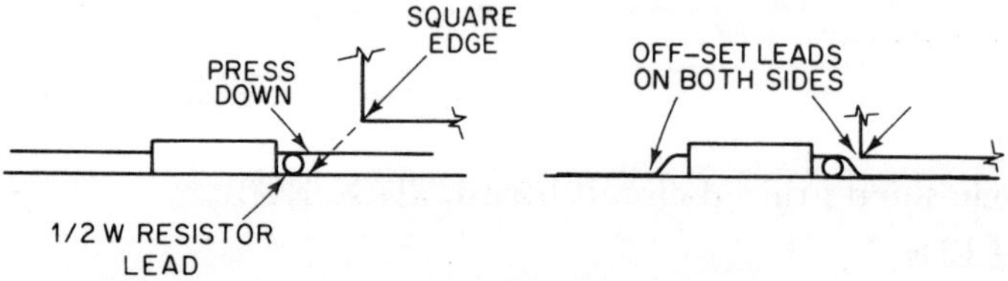

Figure 152 Off-setting flat-pack leads

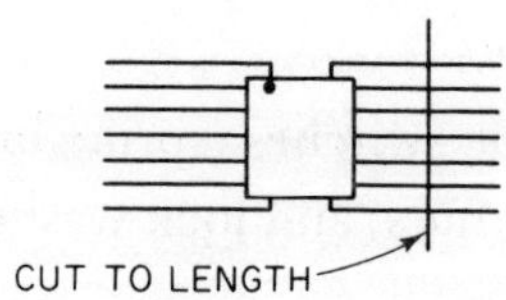

Figure 153 Cutting flat-pack leads

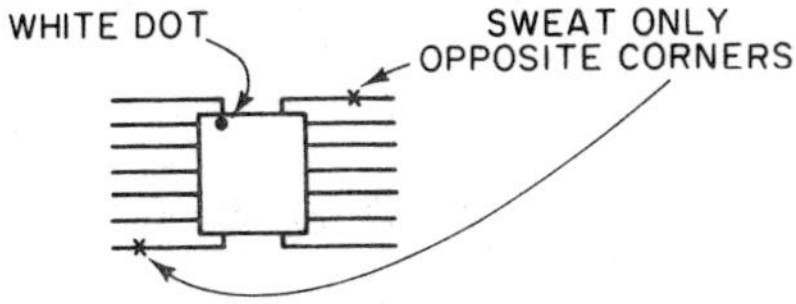

Figure 154 Flat-pack orientation

b. Trim the flat-pack leads to the proper length to fit the pads of the circuit (see Figure 153). Use scissors and cut straight across as indicated.

c. Pre-tin the flat-pack pads on the circuit board. Use plenty of flux and be careful to avoid solder bridging between pads.

d. Place the flat pack on the pad arrangement and sweat opposite corners of the flat pack to the circuit pads. Be sure to orient the white dot of the flat pack to face the pattern dot (see Figure 154).

e. Check to see if the flat pack is straight and perfectly lined up with the pads. If not, loosen one corner at a time and shift to the proper alignment.

f. After the alignment, sweat all leads of the flat pack to the pads. Use plenty of flux. Watch for solder bridging between pads.

4. Construct the switches out of thin spring brass (see Figure 155). Pre-tin the pads and sweat on the switches.

5. Mount the LEDs about 1/8″ from the board, as shown in Figure 156. Orieng the flat side of diode to the face pattern dot.

6. Mount the resistors against the board with the leads cut similar to the LEDs.

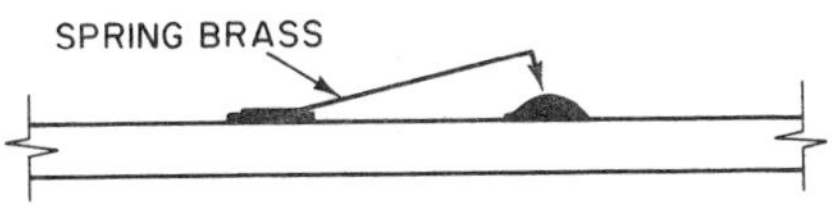

Figure 155 Side view of switches

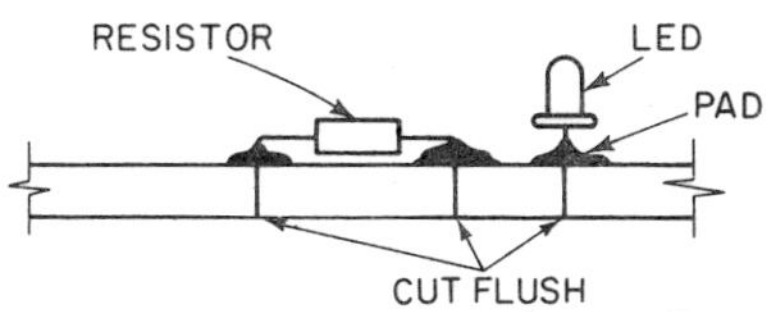

Figure 156 Mounting LEDs

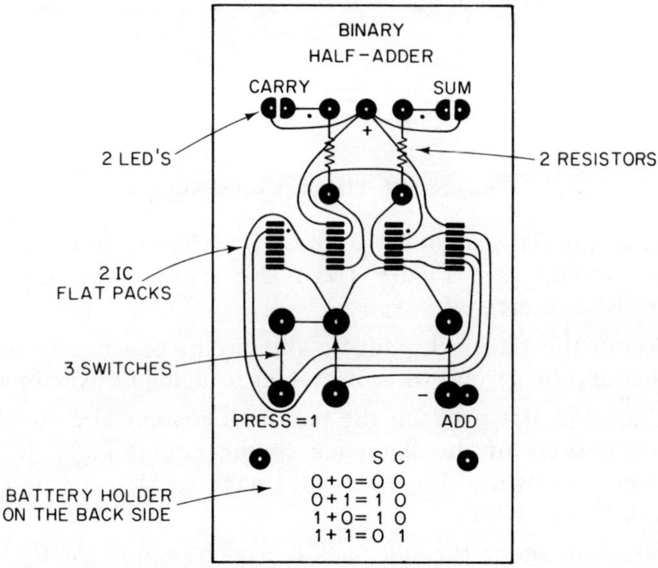

Figure 157 Half-adder layout

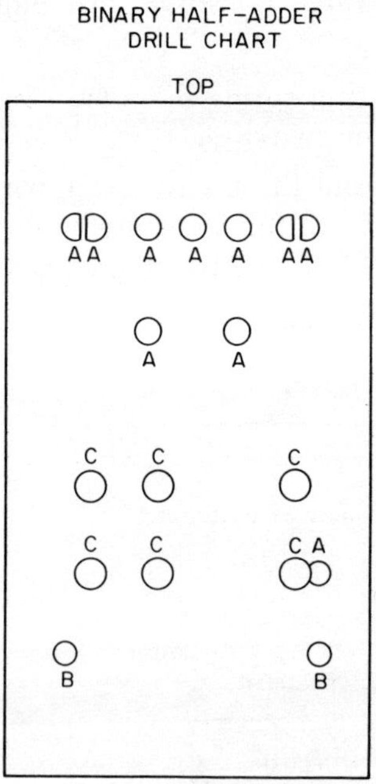

DRILL SCHEDULE

HOLE	DRILL
A	60
B	42
C	NONE

Figure 158 Drill schedule

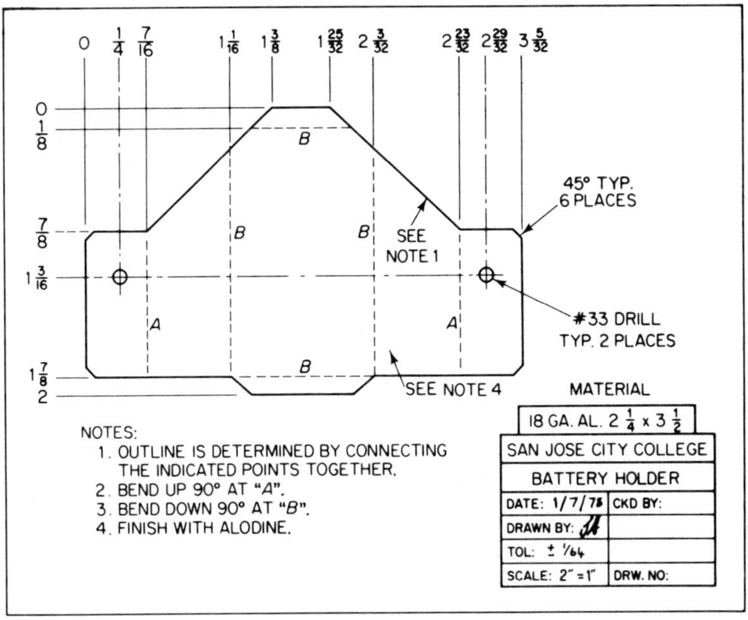

Figure 159 Battery holder

7. Construct the battery holder (refer to Figure 159).
8. Mount the battery holder and battery connector.
9. Test the circuit for proper operation.
10. Use alcohol and the air hose to get the board absolutely clean.
11. Scratch your name on the back of the board behind the word "Binary" and turn it in to your instructor.

TRANSISTOR RADIO (1)

This radio project includes several interesting theoretical concepts relating to electronics, and also a number of assembly or fabrication processes. It is a project adaptable to a person with limited experience in electronics, is straightforward all the way through, and carries with it the anticipation of good results in getting it to operate properly. The project is to be built by the student pretty much from the ground up, with the exception of the earphones and battery. A pictorial view of how the radio might look is shown in Figure 160.

In addition to the assembly techniques incorporated in this radio

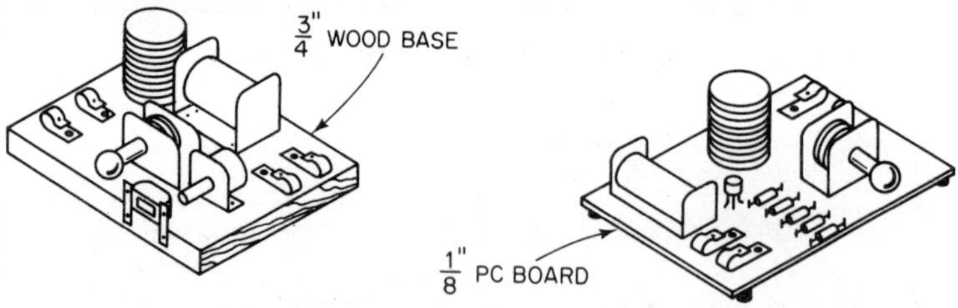

Figure 160 Pictorials of transistor radio

project, there are experiences relating to the use of tuned circuits, amplification, voltage dividers, detectors, transformers, and just simply the use of resistors, capacitors, inductors, and the transistor.

There are three basic radio circuits shown in a progression in Figure 161. The *third* circuit is the one to use in building up this project. The first circuit has detection but lacks in gain and selectivity. In the second circuit there is selectivity and perhaps a small degree of gain as a result of the action of the tuned *LC* tank. Very simple selectivity and gain are built into the third circuit, making it fairly outstanding in performance when it is connected to an outside antenna.

In building up this project, various arrangements can be used, and it is intended that a person's ingenuity be used in working out the problems. In the past it has been built up on a 3/4" wood base requiring point-to-point wiring, and also on a 1/8"-thick printed-circuit board requiring that a printed-circuit layout be designed. Homemade brackets, component mounting facilities, and a homemade coil form are used. Any number of different ideas can be made a part of this project.

Purpose of Assignment

1. To gain experience with assembly or fabrication processes.
2. To gain experience with planning or designing a complete unit.
3. To become acquainted with basic theoretical aspects of a circuit.

Tools and Equipment

Various hand tools.

Soldering device with dampened sponge.

Soldering vise.

Equipment for printed-circuit-board etching and fabrication.

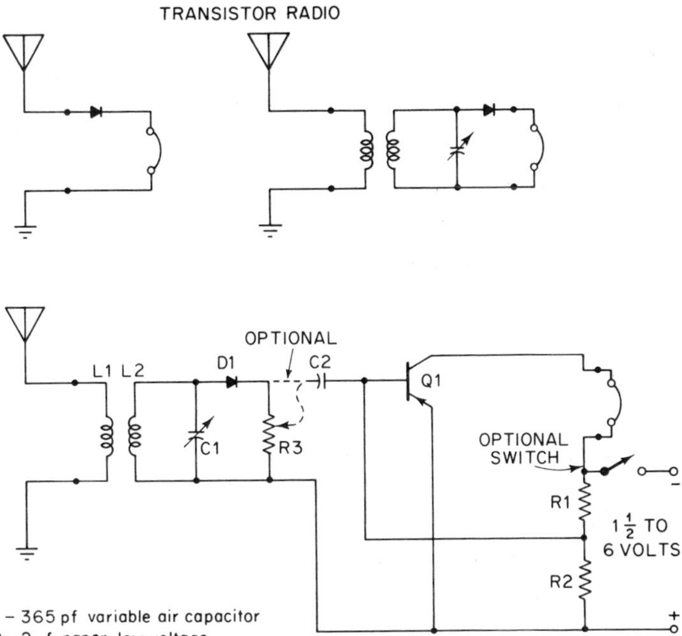

C1 – 365 pf variable air capacitor
C2 – 2μf paper, low voltage
R1 – 68 kΩ, $\frac{1}{2}$ W
R2 – 10 kΩ, $\frac{1}{2}$ W
R3 – 47 kΩ pot., or 47 kΩ resistor, $\frac{1}{2}$ W
L1 – 12 turns no. 25 single cotton-covered magnet wire wound on a 2" cardboard coil form
L2 – 70 turns no. 25 single cotton-covered magnet wire wound on the same coil form as used above. Leave a $\frac{1}{16}$" spacing between coils
2 – Fahnestock clips for antenna and ground
Phone jack for low impedance earphones or use two Fahnestock clips for connecting the earphones
D1 – 1N34 or any similar germanium diode
Q1 – Any high-gain low-frequency transistor can be used

NOTE: 1. The battery can be built onto the chassis in order to eliminate external connections to it.
2. An NPN transistor can be used by reversing the battery connections to the circuit.

Figure 161 Radio circuit schematics

Drill press and bandsaw.
Squaring shear and belt sander.
Box-and-pan brake.
Low-impedance earphones.
Outside antenna.

Materials

Circuit components as listed in Figure 161.

Varnish and quick-drying glue.

2″ cardboard coil form—4″ long.

Sheet metal for making small brackets.

2 X 2″ piece of Masonite.

3/4 X 4 X 6″ wood base or 1/8″-thick printed-circuit board 4 X 5″.

22-gauge rosin-core solder.

1½-V battery.

Miscellaneous screws and bolts.

Assignment

Build a simple transistor radio using the circuit of Figure 161 (refer to the Directions).

Directions

1. Obtain all the components and materials as listed above.
2. Wind *L*1 and *L*2 on the coil form. Keep the turns of wire flat against the coil form and leave a 1/16″ space between *L*1 and *L*2. Secure the ends of the coils and coat the finished transformer with varnish to hold the windings in place (see Figure 162).
3. Cut a circle of Masonite to fit the inside of the coil form and glue it in place (see Figure 162).
4. Decide on whether to use a 3/4 X 4 X 6″ wood base and point-to-point wiring or a printed-circuit-board base.
5. Find a way that all the components could be mounted on the base, which is approximately 4 X 6″ or 4 X 5″. It is suggested that the antenna should connect on the left side and that the battery be mounted in the rear center, the tuning capacitor in the front center, and the earphone connections on the right side. Try different arrangements.
6. If the wooden base is used, cut, sand, and varnish the base.
7. If the printed-circuit-board base is used, study how printed circuits are made (refer to other assignments for information). Make a pencil drawing of a *good* conductor layout for the circuit. Tape up a photomaster, make a 3M Color Key negative, and etch the printed-circuit board.

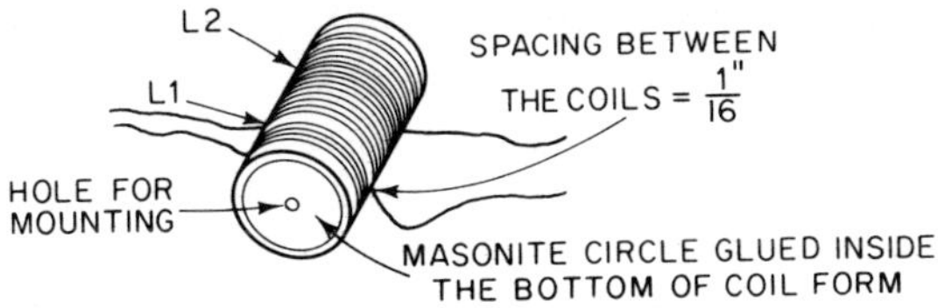

Figure 162 Coil form with mounting base

8. Make mounting brackets, battery holder, and other items needed (see Figure 163).
9. Mount the components on the base and complete all the wiring. Use Fahnestock clips for connecting the antenna and earphones.
10. Connect the completed radio to an outside antenna and low-impedance earphones. Trouble-shoot the circuit as required and study the operation of the circuit.

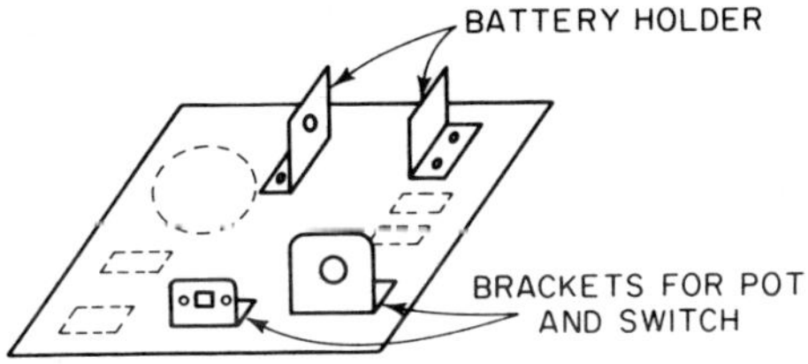

Figure 163 Mounting brackets

LOW-VOLTAGE DC POWER SUPPLY (2)

A simple low-voltage DC power supply probably is one of the most useful devices that a person interested in electronics can own. There can be many uses for this device, such as for bench testing, as a bias supply, for experimentation and design, and as a power source for battery-operated toys and equipment; it can even be used for charging batteries.

The circuit and the instructions in this assignment will enable a person to fabricate and put together a useful DC power supply. The circuit is relatively simple, but includes some desirable features that would go along with this kind of device. Depending upon the AC input, the output voltage is adjustable from 0.7-V DC to approximately 30-V DC and the output is regulated so that it will remain constant over a fairly wide range. In addition, for protection of the circuit, the output current is electrically limited to approximately 0.5 A.

Ordinary standard components are used to make up the circuit so that it almost can be put together out of spare parts. Also, depending upon a person's ingenuity or ability, various alterations can be made to the circuit or the chassis. For instance, the output of the circuit can be metered in terms of voltage only, or in terms of voltage and current. A switching arrangement with two or three functions or ranges also can be used.

Detailed drawings and plans for the construction of this unit are not given in this assignment. It is expected that the student be involved in the detail-planning and decision-making effort that is necessarily a part of prototype work. Certainly, the student's individuality and ingenuity should show in the completed unit and a wealth of experience will be gained.

A suggestion as to how the finished unit might look is shown in Figure 164. The chassis is simple and fairly straightforward in every way, and yet looks amazingly professional if the workmanship is what it should be. The basic arrangement and layout of the chassis shown in Figure 164 is covered in the chassis construction section of the book.

This assignment will involve the student in many different aspects of prototype work. It includes work with evaluation and planning, printed-circuit layout and fabrication, the complete fabrication of a chassis, assembly and wiring, and electrical testing.

Purpose of Assignment

1. To gain experience in evaluating and planning the detail involved in the fabrication of a prototype.
2. To gain experience in the fabrication of the various parts of a prototype unit.
3. To actually end up with a useful low-voltage DC supply for the student to own and use.

Tools and Equipment

Equipment for printed-circuit-board fabrication.

Hand and machine tools for sheet-metal fabrication.

Screen process equipment.

Chemical treatment facilities as required.

Tools for wiring and assembly.

Instrumentation for electrical testing.

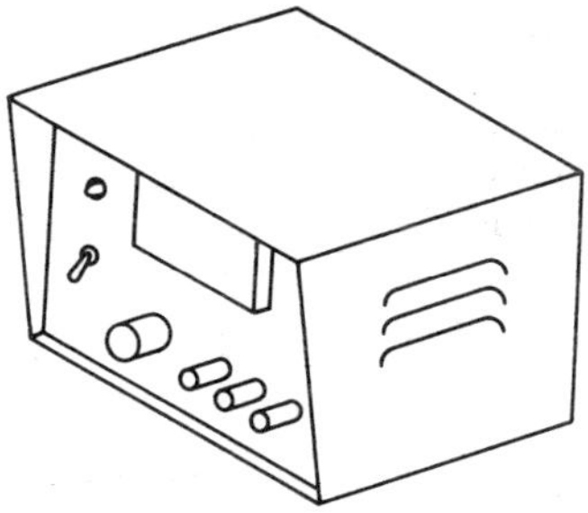

Figure 164 Possible appearance of the DC power supply

Materials

16-gauge and/or 18-gauge sheet aluminum.

Printed-circuit-board material.

Components—see parts list.

Various screws and bolts.

Assignment

Construct a regulated low-voltage DC supply. Determine all of the detail and make plans for fabricating the DC supply unit (refer to Figure 165).

NOTE: This assignment involves determining and making up a set of plans for building a device from the schematic drawing. This planning and problem-solving effort *must be done by the student.* A complete set of detail plans is not included here. Refer to *Layout Development Check List* (Appendix A), *Prototype Development Sequence* (Appendix B), and other appropriate assignments or sections of the appendixes.

Directions

1. Study the circuit drawing to get a basic understanding of how it operates and also what particular components go to make up the circuit. Decide on any alterations that might be made.
2. Breadboard the circuit to check its operation and to become familiar with the position of each part and its relationship to the others.
3. Consider various possibilities of different arrangements or placement of the components in the proposed chassis. For each

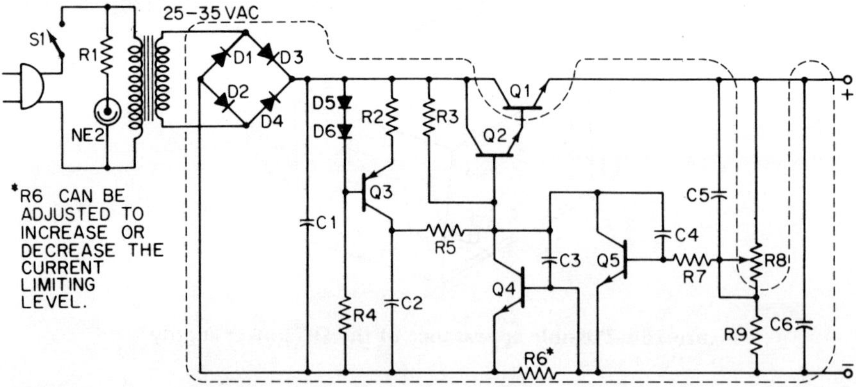

PARTS LIST			
R1	180 kΩ, $\frac{1}{4}$ W	Q1	NPN, 15 W
R2	180 Ω, $\frac{1}{2}$ W	Q3	PNP, Si
R3	4.7 kΩ, $\frac{1}{2}$ W	Q2,Q4,Q5	NPN, Si
R4	10 kΩ, $\frac{1}{2}$ W	T1	25 TO 35 VOLT SEC,
R5,R7	100 Ω, $\frac{1}{2}$ W	S1	SPST
R6	1.6 Ω, 2 W		AC–CAP+CORD
R8	20 kΩ POT.		BANANA JACKS
R9	330 Ω, $\frac{1}{2}$ W		AI FOR CHASSIS
C1	3000 μf, 50 V		BOLTS AND SCREWS
C2,C6	100 μf, 50 V		
C3,C4	300 pF		
C5	10 μf, 50 V		
D1–D4	Si, 1 AMP		
D5,D6	Si		
NE2	Neon		

**MAKE–BEFORE–BREAK SWITCH

***BREAK–BEFORE–MAKE SWITCH

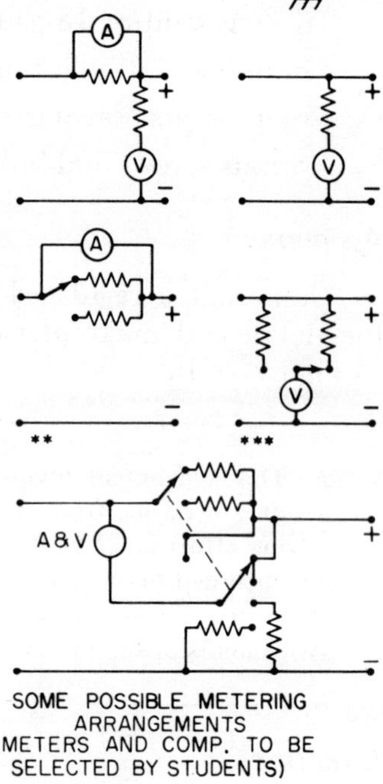

Figure 165 Regulated low-voltage DC supply

component consider the size, weight, heat problems, and mounting problems. Plan to make up a printed-circuit board, including the parts of the circuit that are inside the dotted lines.

4. Make plans with regard to how the series pass transistor, the 20-kΩ potentiometer, and the transformer are mounted. Consider interwiring and connection problems. The series pass

transistor needs to be mounted on a heat sink, and most of the other parts should be on the printed-circuit board. Also, the 20-kΩ potentiometer needs to be mounted on the front panel with some required connections to the printed-circuit board. It is suggested here that it might work best to use turret terminals on the printed-circuit board and hook-up wire to the proper connecting place. As an alternative, wiring can go to a printed-circuit board in the manner shown in Figure 166. Do not just simply solder hook-up wire straight to a printed-circuit pattern.

5. Make some preliminary plans as to how the finished device will look. Make plans with regard to the size and shape of the chassis. Determine what parts go together to make up the complete chassis, whether or not a printed circuit is used, and how all of the parts fit together. Refer to other assignments, especially those relating to chassis layout and fabrication, and those relating to printed-circuit layout work. Make preliminary drawings of the initial plans. Refer to the section on making proper drawings; make the drawings readable and as complete as possible at this point.
6. Reevaluate all the plans carefully with regard to size, location of components, and the interrelationship of all of the components.
7. Decide on the type of material to use in constructing the chassis. Decide what type of finish will be used and how the front panel will be labeled. Refer to other appropriate assignments.
8. Finalize all plans. Ask questions of the instructor as required and obtain his approval before continuing.
9. Make a printed-circuit layout of the part of the circuit that goes on the printed-circuit board. Complete the fabrication of the printed-circuit board and test it for proper electrical operation.
10. Complete the fabrication, labeling, and finishing of all parts of the chassis.

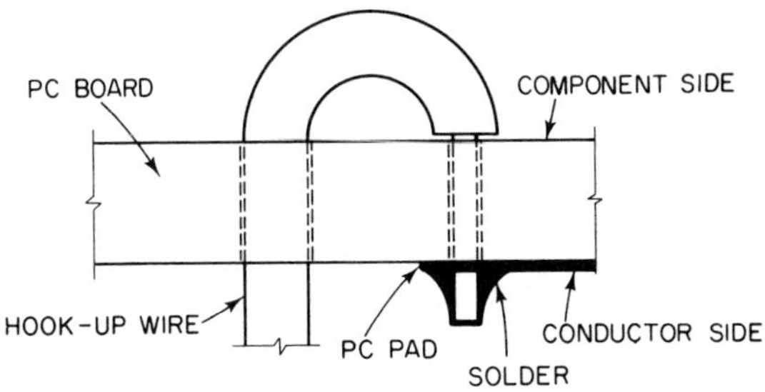

Figure 166 Hook-up wire connecting to a printed-circuit board

11. Completely assemble all the mechanical parts of the unit.
12. Completely wire the unit, with *no* mistakes.
13. Electrically test the completed unit for proper operation. Run tests on input and ouput voltage and current, regulation at different current levels, ripple, and the operation of the current-limiting feature of the circuit. For doing these tests, refer to notes and instructions obtained in electronics theory classes.
14. When the unit is completed, turn it in to your instructor. Be sure that your name is marked clearly on the unit.

PART FIVE

BASIC SHOP TOOLS

9

Thread-Gauge Fabrication

THREAD-GAUGE LAYOUT (1)

Even though an electronics technician should primarily be skilled in the theory and operation of circuits and the use of test instruments, from time to time it is necessary that he have knowledge and skill in the use of certain hand and machine tools. Very often in research and development, in prototype work, and in assembly work, various hand and machine tools are used.

In the next sequence of five assignments, a number of different kinds of operations are involved, and these will require the use of a number of different hand and machine tools. In addition to this, some information and theory on screw threads and twist drills is included in this assignment.

In starting the fabrication of an item, a certain amount of layout work is necessary. The layout could be very simple, or if the item is complicated, the layout work could be quite detailed and require a considerable amount of time. In any case it is wise to give some thought to what is involved and mentally evaluate what some of the problems might be, and the possible solutions to those problems.

Figure 167 gives a review of how dimensions can be accurately transferred from a scale to the workpiece. Notice that the combination square is used against an initially established reference edge and that almost all layout dimensions would be measured from that edge. In some cases two reference edges are used, and in many cases these edges would be related together at a 90° angle, especially with sheet-metal layout work.

Subsequent to the layout requirements on the workpiece, various hand and machine operations would be performed. In performing these operations a definite sequence should usually be followed, so that one particular operation does not interfere with what is to follow.

Before getting too deeply involved in the fabrication of this thread gauge, refer back to the introduction of the *Sheet-Metal Bracket* assignment. The information there has to do with preparing the stock material and making a basic layout.

Purpose of Assignment

1. To gain experience in making a layout.
2. To gain experience in making preliminary plans and following a sequence of operations.
3. To learn how to work carefully and accurately.
4. To learn the importance of double-checking preliminary work before continuing.

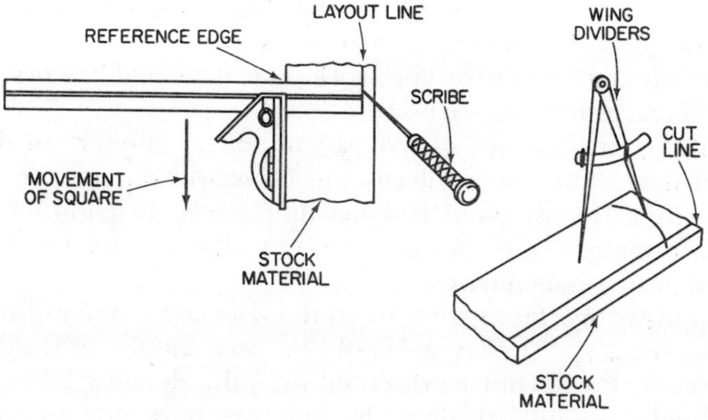

Figure 167 Basic layout techniques

Tools and Equipment

Combination square.

Scribe.

Center punch, anvil, and hammer.

Wing dividers.

Belt sander.

Safety glasses.

Materials

1/4″ aluminum stock, 2 × 5″.

Metal dye, if needed.

Garnet paper, 320 and 600 grit.

Assignment

Make the layout for a thread gauge in preparation for performing subsequent operations. Refer to Figure 168 for the specifications. (Material is 1/4″ aluminum.)

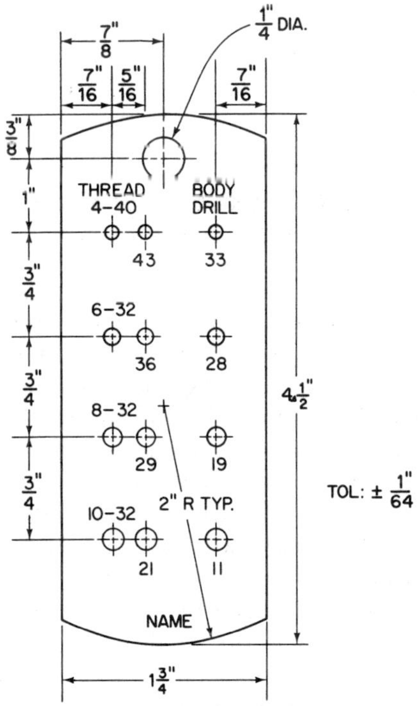

Figure 168 Thread-gauge detail

Directions

1. Refer to the introduction of the *Sheet-Metal Bracket* assignment for information on making a layout.
2. Clean and prepare the surface of the stock material. Inspect the piece to be sure that it will be adequate for making the thread gauge.
3. On the blank stock establish a single straight *reference edge.* The edge should perhaps be on the left and used as a reference for making the layout of the piece, including the cut lines and the location of hole centers. A belt sander could be used for doing this if one is available. *Wear eye protection.*
4. Use the wing dividers for layout of the arcs which describe the ends of the piece. Using the scribe, mark all hole centers and cut lines. Do not scratch the lines all of the way across the piece, only in the area where the lines will be used. Metal dye can be used if desired, especially on places where the scribe line needs to be corrected or moved slightly.
5. With a center punch, *carefully* mark the center of holes to be drilled. Place the workpiece on an anvil when using the center punch.
6. Refer to *Use of Bandsaw* for information on cutting the outline and the directions for subsequent operations.

USE OF THE BANDSAW (1)

(*continued from previous assignment*)

The bandsaw is a machine tool with a motor-driven, endless, belt type of blade. Various materials can be rapidly cut with relative ease and with a very good degree of accuracy. It is not the only answer to all cutting problems but is a good solution on jobs where the material to be cut has a considerable thickness.

Figure 169 shows a typical bandsaw and gives the names of the most important parts. There will be slight variations on different machines but basically these parts will appear as shown in the drawing. Study the indicated items and determine the purpose of each.

Various types of blades can be secured for use in the bandsaw. The type of steel of which the blade is made can vary, as well as the actual structure of the blade. Usually, the blade is made with a *flexible back* and a *hardened* tooth edge and can have various tooth

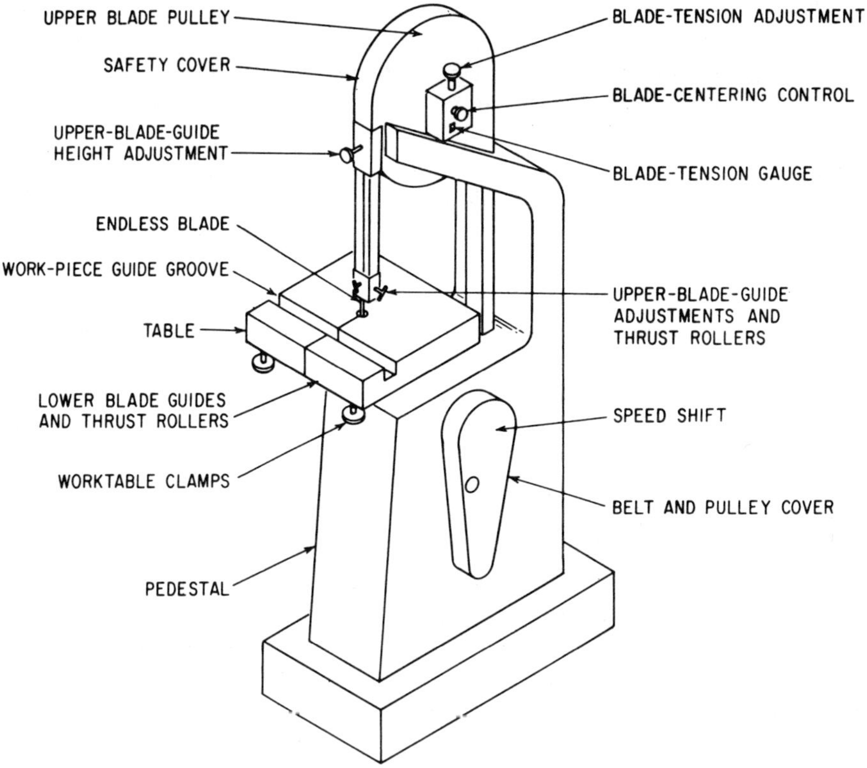

Figure 169 Typical bandsaw

arrangements, such as *hook tooth, wavy set, raker set,* and *skip tooth,* which are shown in Figure 170. The skip tooth is popular and allows for chip clearance. It is used on nonferrous metals, plastic, wood, paper-base materials, and other similar materials. The material is less likely to clog the teeth with this arrangement. The *raker set* is good for mild steel or for contour cutting.

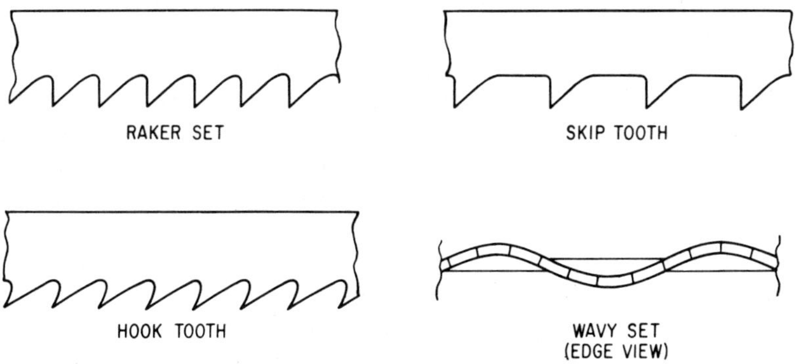

Figure 170 Types of bandsaw teeth

It is recommended that a *high-speed* steel blade (or tungsten or molybdenum blade) be used to extend the use of the saw. If a *carbon-steel* blade is used, only relatively soft materials can be cut; whereas with the harder blades, mild steels can be easily separated with fewer tooth-dulling problems.

The blade stock, which comes in long lengths, needs to be cut to size and butt-welded to make a continuous loop. This welding is done on special jig attachments to the bandsaw. Blade-welding instructions should be included in the factory manual for the saw.

To use the bandsaw properly, quite a number of adjustments need to be made. A centering of the blade as it rides on the wheels is important and is controlled by an adjustment that allows the top wheel to be tilted in or out with respect to the lower wheel. Associated with this adjustment is a tension control, which sets the tightness of the blades. The amount of this tension is determined by the width of the blade itself. This information is indicated by calibrated marks located on the tension adjustment.

The next important adjustment is the setting and positioning of the *upper* and *lower blade guides.* These guides are moved sideways by loosening setscrews, and moved forward and backward by thumbscrew adjustments. Both guides are set to allow the blade to ride free when material is not being cut and to hold the blade in place during the cutting operation.

The teeth of the blade should clear the guides and the back edge should barely rest against the *thrust rollers.* The top guide can be raised or lowered and should be kept fairly close to the work while making a cut.

The worktable on most saws can be tilted a small number of degrees if it is desired. There is usually a "degree-of-tilt" indicator and a clamping handle to facilitate worktable changes. The angle of the table with respect to the blade controls the angle of cut. A sliding workpiece guide is designed to move in a groove on the worktable and also controls the workpiece while the cut is being made.

It is often important to change the speed of the blade. This is done by shifting a belt operating on *cone pulleys,* by shifting a *gear box* or, in some cases, by both methods. It is important *not* to shift the gear box when the saw is turning. Generally, the speed of the blade is determined by the *type of blade* and the *type of material* being cut. For a carbon-steel blade the speed should be slower than for the high-speed blades or the *friction-cutting* type of blade. Also, in cutting hard materials the blade should be slowed down, and it should be increased in speed for the softer materials. Sometimes a coolant is used.

Fine-tooth blades should be used for cutting hard materials or thin material. At least two teeth should engage the work and the feed pressure should be reduced if the work is thin.

The radius of the cut will depend upon the width of the saw blade and to some extent the thickness of the material. As a general rule, if the material is 1″ thick, a 1/4″ blade will be able to cut a 5/8″ radius or a 1/2″ blade will be able to cut a 2¼″ radius. Use a light feed pressure when cutting a radius. For general-purpose work a 1/4″ blade is recommended.

Purpose of Assignment

1. To develop skill in the use of machine tools.
2. To gain experience in the use and function of the bandsaw.
3. To learn safety procedures and develop good working habits.

Tools and Equipment

Bandsaw and goggles for eye protection.

Various hand tools.

File.

Vise.

Belt sander.

Materials

1/4″ aluminum stock, 2 × 5″.

Scrap pieces of aluminum for sawing practice.

Assignment

Lay out the thread gauge shown in Figure 168 on 1/4″ aluminum stock and cut the outline on the bandsaw.

Directions

1. Refer to the *Thread Gauge Layout* assignment for instructions on making the layout of the thread gauge.
2. Inspect the bandsaw to see that it is properly set up. Check the centering and tension on the blade, adjustment of the guides, angle or position of the table, and the drive system.
3. *Wear eye protection.* Turn on the bandsaw and make a few trial

cuts. Determine if the speed of the blade is correct. For soft material such as aluminum the speed of the blade should be relatively fast.

4. Saw carefully and keep your fingers away from the front of the blade. Make sure that the *safety covers* are all in place.
5. Saw the outline of the thread gauge.
6. Use a file to clean up the sawed edge so that it is smooth and straight. Deburr the edges with the file held at an angle. Use a belt sander on the edges of the piece, if available.
7. Refer to the *Use of a Drill Press* assignment for drilling information and the directions for subsequent operations on the thread gauge.
8. Clean up the bandsaw and the surrounding area when finished.

USE OF THE DRILL PRESS (1)

(continued from previous assignment)

Before using a tool a person should be familiar with the nomenclature that pertains to that tool or its parts. Then, it is possible to use the tool more efficiently and with a greater degree of protection for both the operator and the tool. This is especially true with a machine tool such as the drill press.

The basic features or parts of the typical drill press are shown in Figure 171. Proper use of this tool will involve the items that are indicated in the drawing.

An important part of drill-press work is holding the work properly while it is being drilled. The safety of the operator, protection of the work, and the prevention of drill breakage are contingent upon securing the work rigidly in place. This is usually done by one of the following methods:

1. Clamp the work to the drill table with a *C clamp* or special clamps.
2. Clamp the work in a *drill vise,* which in turn is secured to the drill table with clamping bolts.
3. *Use special jigs* for holding the work.
4. Long pieces can be hand-held provided there are no sharp edges that can cut the hand in case of a slip.

Always ensure that the drill will *not* cut into the drill table, the

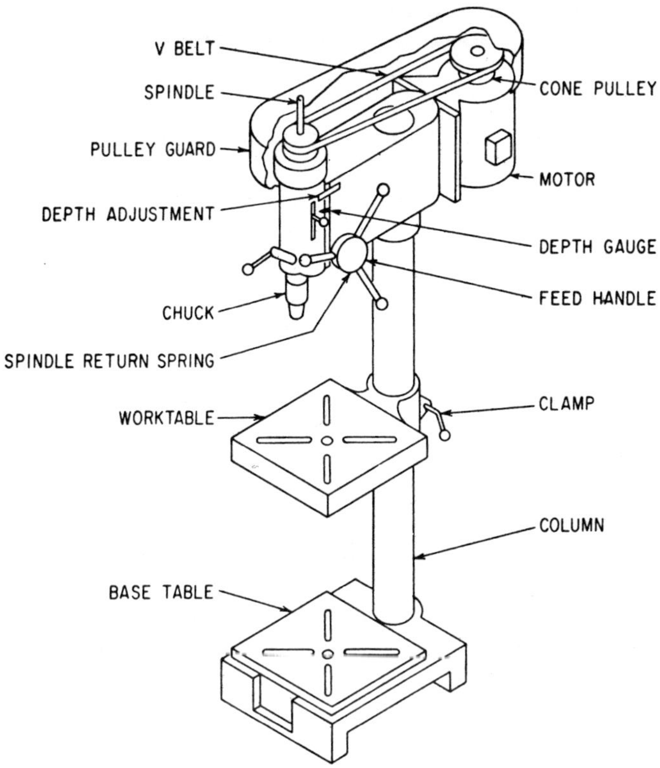

Figure 171 Drill press

drill vise, or any jigs by using a piece of wood between the work and whatever it is resting upon.

The *speed of the drill* is another important factor to consider when drilling. Formulas can be used for making calculations, and there is a drill-speed chart in Appendix C, which can be referred to for this information. However, the drill speed is generally dependent upon the *size of the drill* and the *type of material* being drilled. For large drills the drill speed should be *slow,* and for soft materials the drill speed should be increased. This information is presented in the drill-speed chart.

Speed changes for the drill are made by changing the position of the V-belt on the cone pulleys.

A lubricant is often used when drilling, especially on a fast production line when the drills are in constant use. A chart listing the various lubricant requirements is given in Appendix C.

It should be noted in the chart that a lubricant is not always needed when drilling many of the materials listed, and actually should not be used at times. In actual practice in the small shop, the use of a lubricant is quite limited due to the untidiness involved. Therefore, its use is quite often left strictly up to the judgment of the operator.

It is recommended by the author that only good-quality *high-speed drills* be used. The carbon-steel drills are relatively soft and will not hold a sharp cutting edge. With the use of high-speed drills it is not difficult to always have a sharp drill on hand.

In addition to the ordinary, general-purpose twist-drill point, a drill can be sharpened in a special way in order for it to serve as a *sheet-metal drill.* This is shown in Figure 172.

An ordinary twist drill should not be used for drilling thin sheet metal, because it tends to catch as it breaks through. Also, it will not form a round hole but, instead, one that is triangular. If at all possible, it is best to use a turret punch for punching holes in sheet metal rather than to drill them.

Quite often when drilling a large hole through a thick piece of material, a guide hole is drilled first by marking the center and then drilling with a fairly small drill. This will prevent the larger drill from "running" or drifting.

Be sure to use eye protection when drilling, and also, to protect your hands from being cut, be sure to properly clamp the piece that is to be drilled.

Purpose of Assignment

1. To develop skill in the use of machine tools.
2. To gain experience in the use and function of the drill press.
3. To learn safety procedures and develop good working habits.

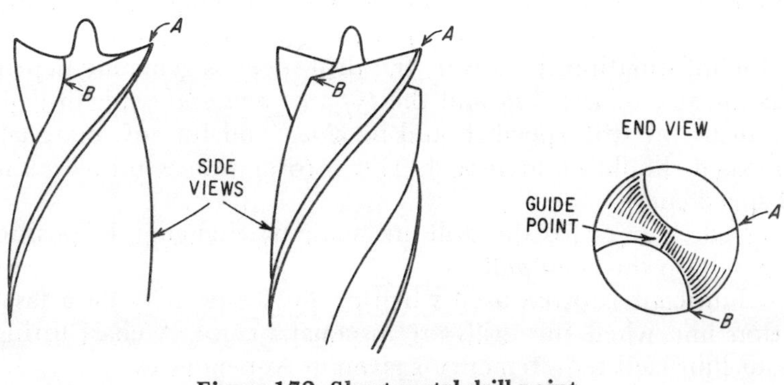

Figure 172 Sheet-metal drill point

Tools and Equipment

Drill press and goggles for eye protection.

C clamp.

Drill vise.

Twist drills, regular and special ground.

Center punch, anvil, and hammer.

Various hand tools.

Materials

Scrap steel for drilling practice.

16-gauge aluminum sheet (approx. 2 X 2") for drilling practice.

1/4" aluminum stock, 2 X 5" (refer to *Thread-Gauge Layout*).

Assignment

1. Practice drilling sheet aluminum and thicker pieces of mild steel. Practice controlling the "breakthrough" of the drill.
2. Locate and drill the holes required in the thread gauge as shown in Figure 173. Follow the directions given below.

Directions

1. Lay out the thread gauge on 1/4" aluminum following the information given in the *Thread-Gauge Layout* assignment.
2. Cut out the thread gauge. Information is given in *Use of the Bandsaw*.
3. Choose the proper-sized drills from the information given in Figure 173. Refer to the body- and tap-drill charts in Appendix C.
4. With a center punch, mark the center of the holes to be drilled. Read the introduction of this assignment and drill the holes carefully. *Wear eye protection.*
5. Refer to *Use of the Thread Tap* before tapping the threads.
6. Refer to *Letter and Number Stamping* for marking the thread gauge.
7. When completed, turn in the work (with your name stamped on it) to your instructor.
8. Clean up after you have finished drilling.

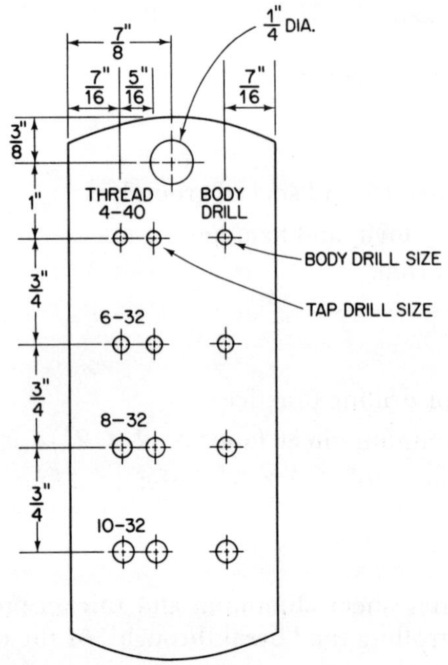

Figure 173 Drill-gauge hold detail (*Note:* size of holes to be determined by student)

USE OF THE THREAD TAP (1)

(continued from previous assignment)

The thread tap is a precision hand tool used for the purpose of cutting machine screw threads in metal. These threads are designed to receive a bolt or screw of a particular size and usually are similar to the type of a thread that is found on the inside of a nut.

The thread most commonly cut by a tap is the *American National Form* thread, which is in some ways similar to the V-thread. A profile of this thread and an enlargement of the detail are both shown in Figure 174.

Currently there are five classes of this thread in use: the *National Fine* thread (NF), which formerly was called the SAE thread; the *National Coarse* thread (NC), which was formerly called the U.S. Standard thread; and three classes (N-8, N-12, N-16), which are quite specialized as to pitch and are used mainly where large diameters are involved, such as in a packing gland, an adjusting collar, or a retaining nut.

Refer to Appendix C for more information on various thread types.

As indicated in Figure 174, a machine screw bolt has a *minor* and a *major* diameter. The minor diameter is measured at the bottom (root) of the threads, and the major diameter is measured at the top (crest) of the threads. In drilling a *tap hole* or a *body clearance hole* the size of the drill is *approximately* equal to these dimensions.

To provide adequate clearance for the crest and root, the design of the internally tapped thread is modified to vary slightly from the bolt thread. This is shown in Figure 174. The crest of the *tap* threads extends the major diameter of the nut as shown at *J*. Also, proper choice of the *tap drill* increases the minor diameter of the nut as indicated at *K*.

The tap-drill size is figured on the basis of allowing the tapped thread to cover approximately 75% of the bolt thread. It has been determined that use of the top edge (crest) of the thread does not materially increase its holding strength. Also, in practice, the thread-

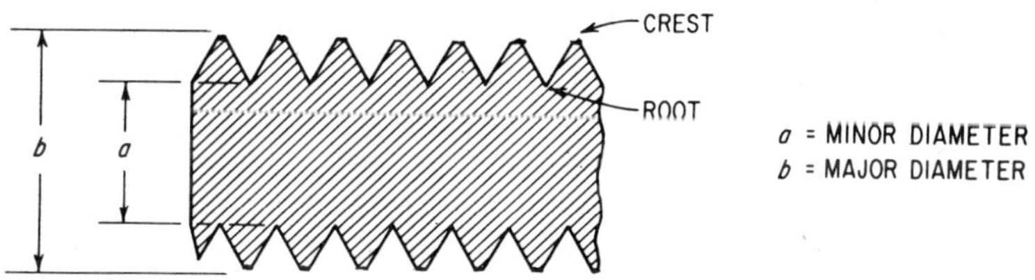

Figure 174(a) American National Form thread profile

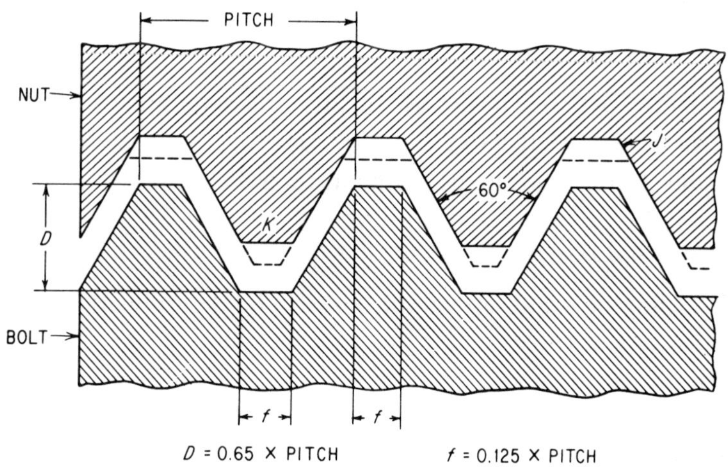

Figure 174(b) Nut and bolt profile (enlarged view)

cutting operation is made simpler by not having to form a fine edge on the top of the thread. Tap-drill-size information can be obtained from charts such as found in Appendix C or can be calculated very closely by use of the following formula:

$$\text{tap-drill size in inches} = \text{body size} - \frac{1}{\text{threads per inch}}$$

The American National Form thread is designated by use of a four-term expression. For example, the expressions 10-24NC-2 and 1/4-28NF-3 give the required information regarding a thread. These expressions are interpreted as follows:

1. The *first term* indicates the *size*. This can be a whole number or a fraction.
2. The *second term* indicates the number of *threads per inch* counted by measuring from the top of one thread to the top of a succeeding thread.
3. The *letters* NC, NF, and N tell what *class* of thread is considered.
4. The *fourth term* gives the class of *fit* of the thread using the numbers from 1 to 4, where 4 is the tightest or closest tolerance.

The *taper tap, plug tap,* and the *bottoming tap* are three basic types of tap. These are shown in Figure 175. These types differ mainly in the amount of chamfer. The *taper tap* should be used for normal work and the other two for the purposes indicated by their names.

Purpose of Assignment

1. To gain experience in the use of the threading tap.
2. To gain a knowledge of screw threads.
3. To develop skill in the use of hand tools.

Tools and Equipment

Threading taps and tap handle, sizes to be determined.

Various hand tools.

Drill press and goggles.

Vise.

Thread gauge.

Twist drills, sizes to be determined.

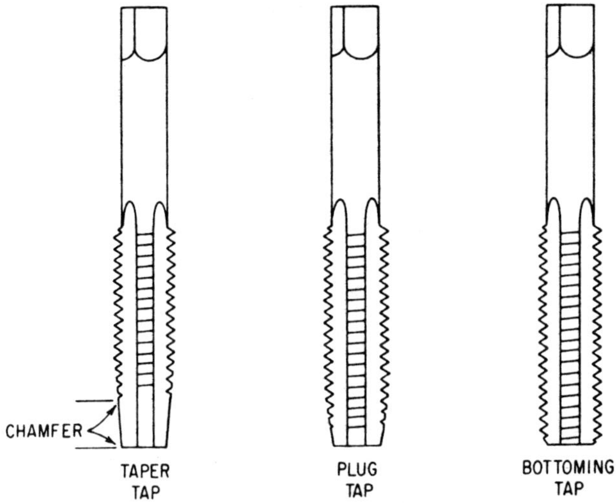

Figure 175 Types to taps

Materials

1/4″ aluminum stock, 2 × 5″.

Threading oil (lard oil).

Assignment

Tap the threads in the *thread gauge* as indicated in Figure 176.

Directions

1. Refer to the Directions in *Use of the Drill Press.*
2. Mount the thread gauge horizontally in a vise, being careful not to mar the edges of the aluminum piece.
3. Select the proper thread taps for tapping the holes.
4. Tap the holes. Start the tap carefully, making sure that it is always perpendicular to the work.
5. After one or two threads have started to "take hold," reverse the tap approximately one-third turn. This breaks the chips loose and clears the tap.
6. *Apply plenty of oil.* Turn the tap in the forward direction approximately two-thirds of a turn. Reverse the tap one-third of a turn to break the chips loose. Continue turning the tap *forward* approximately two-thirds of a turn and then *reversing* one-third of a turn.

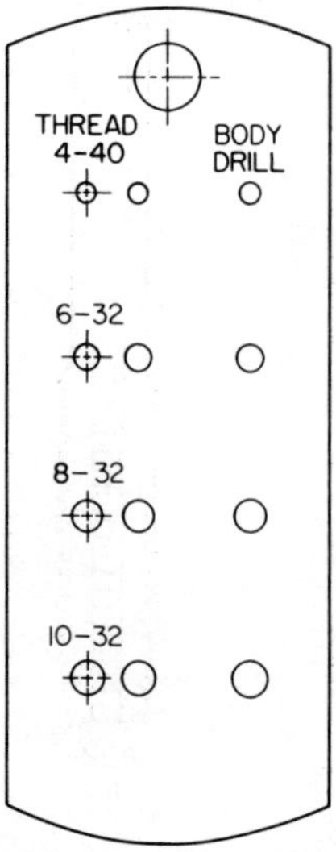

Figure 176 Thread sizes to be tapped

7. Do not apply too much pressure to the tap at any time.
8. Continue until the threads are complete.
9. Refer to *Letter and Number Stamping* for marking the thread gauge with your name and other required labeling.

LETTER AND NUMBER STAMPING (1)

(*continued from previous assignment*)

Letter and number stamps are made for the purpose of marking the identification of controls, parts, and other types of labeling. These stamps are made of a hardened steel and can easily make a letter or number depression in soft metals, various materials, and mild steel. The letters or numbers come in different sizes and are formed onto a square shank, which is held by hand and tapped with a hammer to make the impression.

The advantage of using this method of labeling is that it is quick and easy and makes a permanent impression, which cannot easily be removed.

The quality of the numbers and letters is such that this type of labeling usually would *not* be used on a front panel, where appearance is important. The definition of the characters is not always sharp and clear and it is difficult to keep them straight and even.

In addition, the characters are swaged into the metal and in some cases will distort the metal or cause a defect to appear on the back side. This is mainly true with stampings on sheet metal. However, with a small degree of care, a fairly neat job of labeling can be accomplished in certain applications.

As mentioned above, one of the problems associated with the use of these stamps is that of keeping the characters in a straight and even line. As shown in Figure 177, the use of a *guide block* will alleviate this problem to a great extent. This block is held rigidly in place with C clamps. Be sure that the piece being stamped is backed by a heavy, solid piece such as a flat stake or a heavy piece of smooth iron.

Purpose of Assignment

1. To show one method of marking and labeling for identification.
2. To gain experience in working cautiously and carefully.

Tools and Equipment

Block or straightedge to use as a guide.

C clamps.

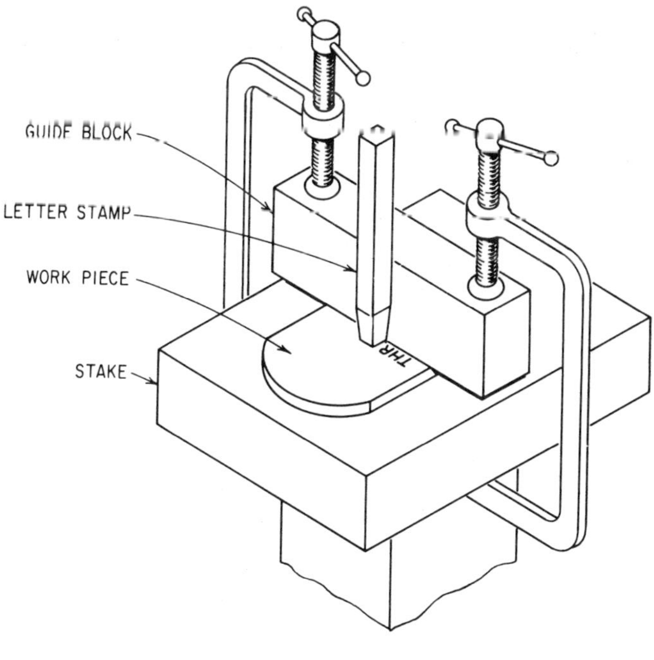

Figure 177 Method of using guide block

Ball-peen hammer.
Flat stake.

Materials

Scrap pieces for practice.
Workpiece to be marked or labeled.

Assignment

Choose an item that needs to be marked. This can be the thread gauge suggested in *Use of the Drill Press* and shown in Figure 178.

Directions

1. Using the guide clamped to a piece of scrap metal, practice stamping a few letters in an even line.
2. Tap the letter or number stamps firmly with the hammer. Make all the letters and numbers appear to have the same depth. The period, dash, and letters similar to the *I* should be struck with a lighter force. Check the distance between the letters.
3. Determine the location and position of the labeling required on the thread gauge shown in Figure 178.
4. Clamp the thread gauge and guide on the flat stake with the

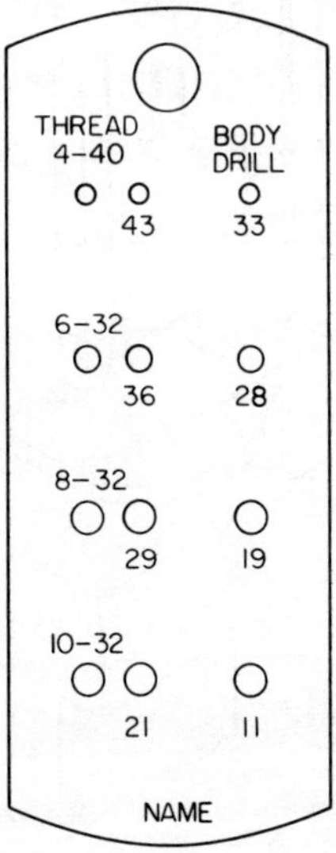

Figure 178 Thread-gauge lettering detail

guide in the proper position for a particular line of letters or numbers (refer to Figure 177).

5. Stamp the letters or numbers as required. Check all letters and numbers carefully. If necessary, some of the characters can be restamped by locating the proper stamp in the original depression and tapping again with the hammer.
6. Make *sure* that none of the characters are accidentally stamped upside down.
7. When finished, clean up the thread gauge or other stamped item and turn it in to your instructor for evaluation.

10

Tools and Tool Sharpening

USE OF THE THREADING DIE (1)

The *threading die* is a hand tool used for the purpose of cutting external threads on round stock, such as a bolt shank or shaft. Usually, the type of thread that is cut by the die is the *American National Form* thread and is designed to fit the *National Fine* or *National Coarse* thread in the ordinary nut. Refer to the introduction of the *Use of the Thread Tap* assignment for a description of the profile of this thread and a discussion of the designations as to pitch, size, type of thread, and class of fit.

The threading die itself is usually held in a handle commonly referred to as the *die stock.* A drawing of this die stock is shown in Figure 179.

In this drawing notice the *handles* used for hand-turning the die and the *clamping screw* used for holding the die in place. The die itself is replaceable, depending upon the choice of thread that is to be cut, and can be either adjustable or fixed.

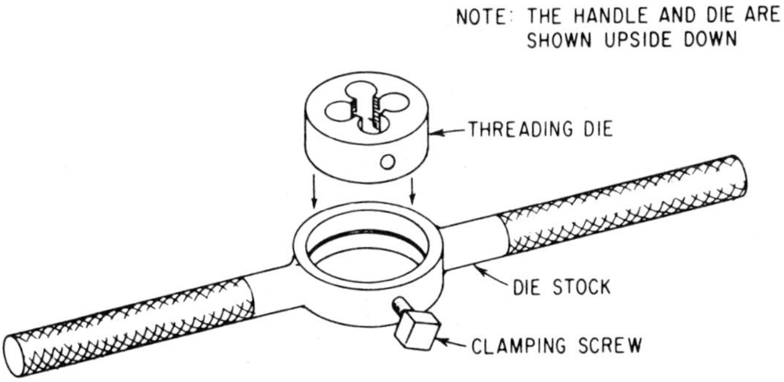

Figure 179 Threading die and holder

Purpose of Assignment

1. To develop skill in the use of hand tools.
2. To gain experience in the use of the threading die.
3. To gain a knowledge of screw threads.

Tools and Equipment

Threading dies and handle sizes to be determined.

Various hand tools.

Vise.

Hand file.

Materials

Round stock, mild steel 1/2″ diameter × 3″.

Round stock, mild steel 1/4″ diameter × 3″.

Threading oil.

Assignment

Select pieces of round stock (mild steel) for practice in making external threads with a threading die. Choose the proper size dies and follow the instructions given in the Directions.

Directions

1. Refer to *Use of the Thread Tap* and read the introduction.
2. Chamfer (to a 45° angle as shown in Figure 180) the end of the round stock that is to be threaded. Mount the stock vertically in a vise.

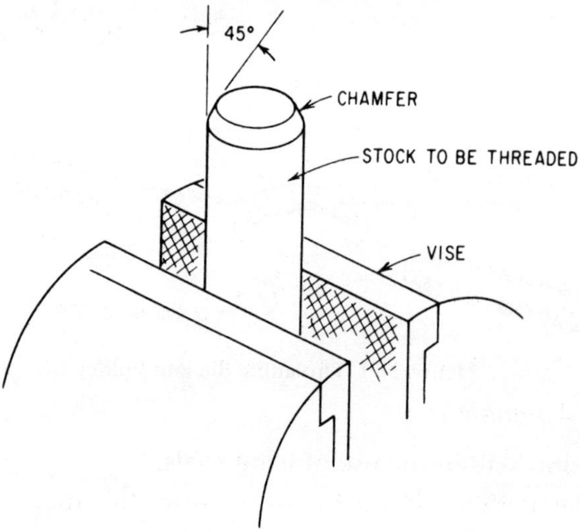

Figure 180 Round stock ready for threading

3. Select the proper dies for threading the round stock that is chosen.
4. Start the die carefully, making sure that it is always perpendicular to the work. Apply a small amount of cutting oil (lard oil).
5. After one or two threads have started to "take hold," reverse the die approximately one-third of a turn. This breaks the chips loose and clears the die for further cutting.
6. *Apply plenty of oil.* Turn the die in the forward direction approximately two-thirds of a turn. Reverse the die one-third of a turn to break the chips loose. Continue turning the die *forward* approximately two-thirds of a turn and then *reversing* one-third of a turn.
7. Do *not* apply too much pressure to the die at any time.
8. Continue until the threads are complete. Have the instructor check the threads and also run a nut on the threads to see how it fits.

ANGLE GRINDING NO. 1 (1)

In a shop where the grinder is used primarily for tool sharpening, the grinder must be a *precision* tool, and treated as such in all respects.

The grinding wheel is composed of silicate, carborundum, or aluminum oxide, which are very hard substances. The structure of the wheel is such that it is porous, with sharp cutting edges of the material formed around the pores. Proper care of the wheel should be as follows:

1. Use the wheel dresser to sharpen the cutting edges of the structure.
2. Use the wheel dresser to true up the grinding surface.
3. Prevent the wheel from becoming clogged with soft materials such as aluminum, copper, wood, and plastics.
4. Watch for cracks and wheel unbalance.
5. Properly adjust the work rest and the wheel guards.

It should further be pointed out that in precision grinding, normally a very small amount of material is removed. This necessitates holding the work very lightly and accurately against the wheel.

Purpose of Assignment

1. To introduce the grinding wheel and the basic techniques of its use.
2. To gain skill in the use of the grinding wheel for grinding an angle to a very close tolerance.
3. To gain background experience in preparation for tool grinding.

Tools and Equipment

Grinder.

Wheel dresser.

Goggles or face shield.

Angle gauge.

Hacksaw.

Cooling water.

Materials

1/8 × 1/2″ strap iron approximately 4″ long.

Assignment

Grind the angle as shown in Figure 181.

Directions

Before grinding, check the grinding wheel for flatness and cleanness. If the surface of the wheel is not right, use the wheel dresser

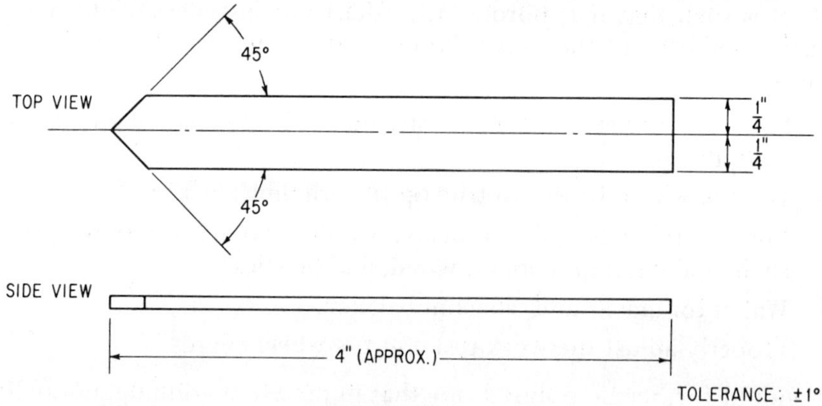

Figure 181 Angle-grind detail

to correct it. *Wear eye protection.* Check the work rest for proper adjustment.

1. Cut the basic point angle with the hacksaw. There is no particular need to grind off these corners. It would be quite hard on the surface of the wheel.
2. Use the coarse grinding wheel for truing up the basic point angle.
3. Cool the work at intervals while grinding.
4. Use the fine wheel for refining the point angle to the exact number of degrees. Grind very slowly and accurately. The angle is out of tolerance if the least bit of discrepancy can be detected when checking it.
5. Check your results with the instructor.
6. Deburr edges of the grind very, very lightly with the fine grinding wheel. *Do not* use a file for this.
7. Stamp your name on the piece and turn it in to your instructor.

SHARPENING CENTER PUNCH, SCREWDRIVER, AND SCRIBE (1)

To do good work, tools must be *sharp.* This is true not only with *cutting* tools but also with other types of tools, such as the screwdriver or a *marking tool* like the center punch or scribe. The screwdriver or wrench will not hold well if the contact surfaces are rounded. The center punch will not make a clean mark and the scribe will not make a fine clean line if these tools do not have a good point.

The sharpening operation is a precision operation and involves a considerable amount of carefulness on the part of the technician. It is important that the grinding wheel be smooth and in good condition. Also, it is essential to be aware of the fact that in most sharpening operations only a *very small amount* of material is removed from the tool. Material is removed to make the edges sharp or surfaces smooth, and this is accomplished by grinding *only in the right place.*

The various angles or shapes that are formed onto the center punch, screwdriver, and scribe are shown in the directions. The center punch and scribe essentially are made pointed and the screwdriver is sharpened to have square edges. The face of the side of the grinding wheel can be used for establishment of these surfaces. Refer to the *Angle-Grinding* assignments for information on the use of the grinder.

The above mentioned tools are hardened or tempered on the point. It is essential that the point of the tool *not be heated* while grinding. Excessive temperature would destroy this temper.

Purpose of Assignment

1. To learn how to sharpen the center punch, screwdriver, and scribe.
2. To gain experience in the use of tools.
3. To gain experience in working carefully and accurately.

Tools and Equipment

Grinding wheel.

Cooling water.

Eye-protection goggles.

Angle gauge.

Wheel dresser.

Oil stone.

Materials

Dull center punch, screwdriver, or scribe.

Assignment

Sharpen the center punch, screwdriver, or scribe.

Directions

1. Check the fine grinding-wheel surface for flatness and sharpness (cleanness). Use the wheel dresser to correct it if it is not right. *Wear eye protection.*
2. Check the work rest for proper adjustment (close to the grinding wheel).
3. When sharpening the screwdriver, hold it against the wheel *very lightly* at the proper angle. While continuing to grind, check often to see that the wheel is grinding in the right place. Use cooling water at intervals. The sides of the screwdriver blade should be nearly parallel (refer to Figure 182).

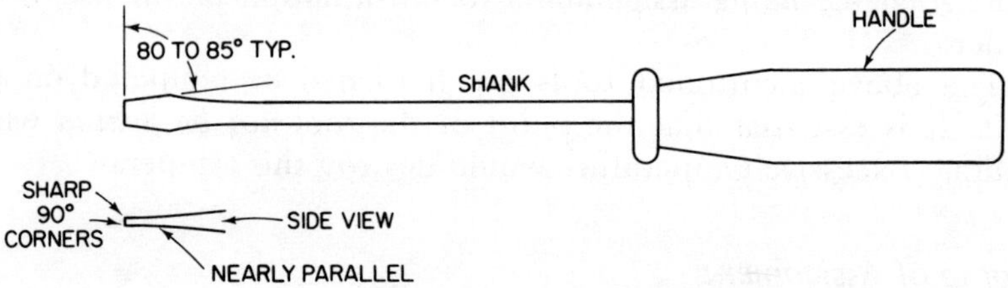

Figure 182 Screwdriver blade angles

4. When sharpening the center punch, hold it against the wheel *at the proper angle* and *rotate* while grinding. Use one hand resting on the work rest as a guide and the other hand to rotate the tool. Grind very slowly and carefully. Use cooling water at intervals (refer to Figure 183).

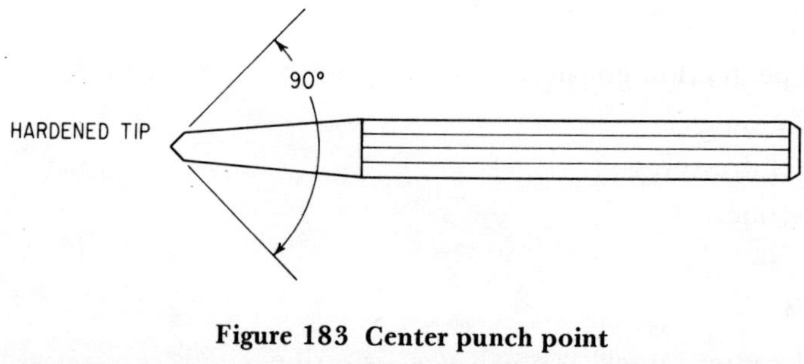

Figure 183 Center punch point

5. Before sharpening the *scribe,* determine whether the point can be corrected by use of the oil stone only. If not, then grind *very lightly* on the fine wheel in much the same manner as when grinding the center punch. Use cooling water and do not let the

tempered point get hot. If the grinding wheel is used, the end point would then be perfected by using the oil stone. This is actually a delicate operation and great care *must* be used (refer to Figure 184).

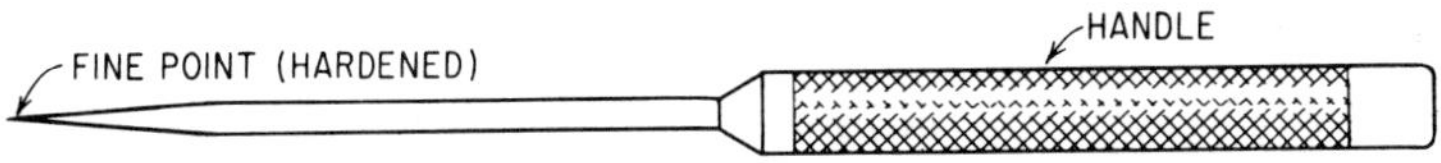

Figure 184 Sharpening a scribe

6. Check your results with your instructor.

ANGLE GRINDING NO. 2 (2)

In tool sharpening, it is important to be able to grind a small surface to a very accurate angle. It is therefore necessary to have good control of the tool or workpiece as it is held against the grinding wheel. It is essential to realize that in most tool-sharpening operations, a very, very small amount of metal is removed.

The sharpening process is simply a delicate refinement of the cutting-edge sharpness. There are basically two starting points in the tool-sharpening sequence.

1. Establish the basic angle and then sharpen the tool to its finished form. (Broken or chipped twist-drill bits or lathe-tool grinding are examples.)
2. Work from the established angle and merely refine the sharpness of the tool's cutting edge. (A screwdriver, twist drill, tin snips, and lathe-tool sharpening are examples.)

Purpose of Assignment

1. To gain additional experience in the use of the grinding wheel in preparation for tool sharpening.
2. To gain experience in grinding a surface measured in two directions.
3. To gain experience in working very carefully and accurately.

Tools and Equipment

Grinding wheel.

Wheel dresser.

Goggles or face shield.

Angle gauge.

Cooling water.

Materials

1/4 X 1″ strap iron, cold-rolled, approximately 4″ long.

Assignment

Grind the end of a piece of stock to the angles, as shown in Figure 185.

Directions

Check the grinding wheel for the proper grinding surface. If the surface is not right, correct it with the wheel dresser. *Wear eye protection.* Check the work rest for proper adjustment.

1. Establish the basic point angle with the coarse grinding wheel. Cool the work while grinding.
2. With the fine wheel, refine the point angle to an exact 45° and an exact squareness with the edge. The angles are out of tolerance if you can see that they are the least bit off from the prescribed angles.
3. Check to be sure that the vertex of the point is in the exact center of the piece.
4. Check your results with the instructor.

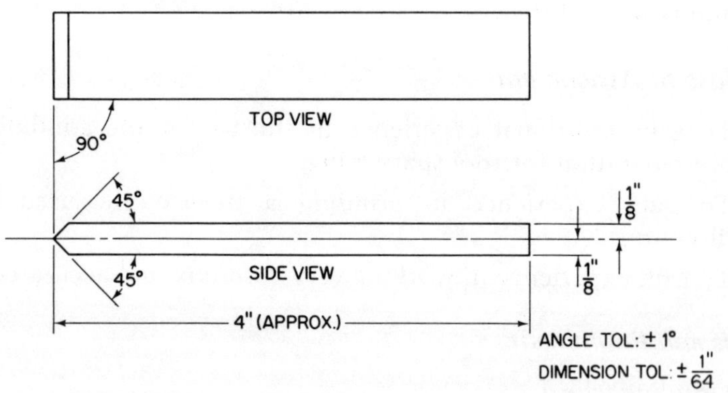

Figure 185 Angle-grinding detail

5. Deburr the edges of the grind very slightly with the fine grinding wheel.
6. Stamp your name on the piece and turn it in to your instructor.

TWIST-DRILL POINT GRINDING (2)

Tools must be sharp if one is to do good work. It is considerably harder to work with cutting tools that are not sharp, and poor quality is the result. This is true with many tools, including the twist drill.

The lip-relief angle, point angle, cutting lips, and the chisel edge (Figure 186) are four important factors that need to be considered in twist-drill point grinding. Depending upon the angle at which the drill is held against the grinding wheel, all four of these angles or edges are *simultaneously* established during the initial part of the sharpening operation.

Essentially, the point-grinding operation includes two steps:

1. The initial grind (which establishes the four angles or edges mentioned above).
2. A slight turn of the drill and then a very, very slight grind on the heel (which increases the lip-relief angle).

Purpose of Assignment

1. To learn how to properly sharpen a twist drill.
2. To learn the essential twist-drill point angles.
3. To gain experience in grinding to accurate and close tolerance.

Tools and Equipment

Twist-drill gauge.

Grinder and goggles.

Scale with 1/64″ divisions.

Magnifying glass.

Wheel dresser.

Materials

3/8″, 7/16″, or 1/2″ twist drill.

Water for cooling.

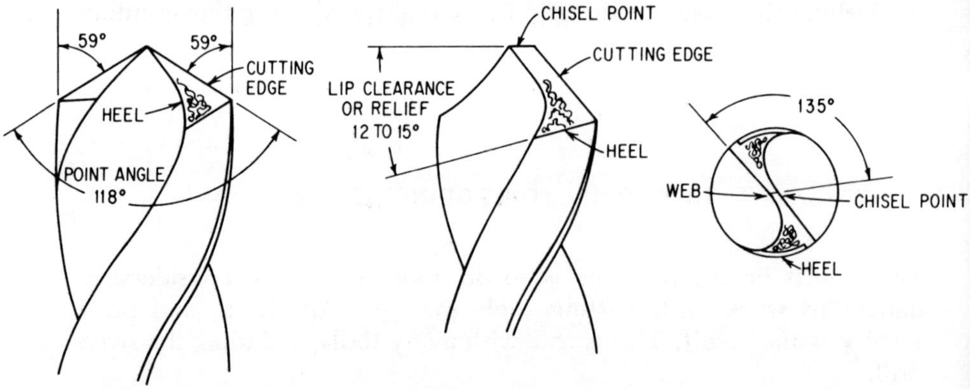

Figure 186 Twist-drill point angles

Assignment

Sharpen a twist drill, following the instructions given in the Directions.

Directions

1. Hold the drill against the wheel with the cutting-lip edge horizontal and pointing up. The body of the drill slants down at a 12 to 15° angle, and to the left at approximately a 30° angle. This position *simultaneously* establishes the lip-relief angle, point angle, cutting lips, and the chisel edge (see Figure 187). *Wear eye protection.*
2. Adjust the angle of the drill against the face of the wheel to establish correct lip relief and point angle.
3. Grind *very, very* lightly. Keep the drill cool.
4. Do not let the chisel edge get off center.
5. Keep turning the drill 180°, alternately grinding one face, then the other, and continuously check the angles and edges for correctness.
6. Measure the edges of the cutting lips using a magnifying glass to obtain a high degree of accuracy. These *must* be the same length (refer to Figure 188).
7. Measure the cutting-edge angles with a high degree of accuracy. They *must be the same* on both sides. It is more important that they be the same rather than any particular angle (refer to Figure 188).

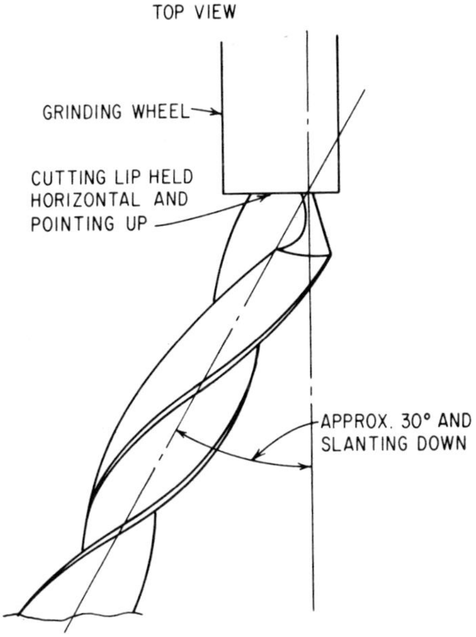

Figure 187 Sharpening a twist-drill point

8. Very slightly turn the drill to lift the *cutting-lip edge* and the *chisel edge* away from the grinding wheel and very lightly grind the *heel* (Figure 186) to increase the *lip-relief* angle.
9. Continuously check the drill point while grinding. Keep the drill *cool.*
10. When finished, inspect the drill point for chips or cracks and for sharpness.
11. Try out the drill to see how it works.

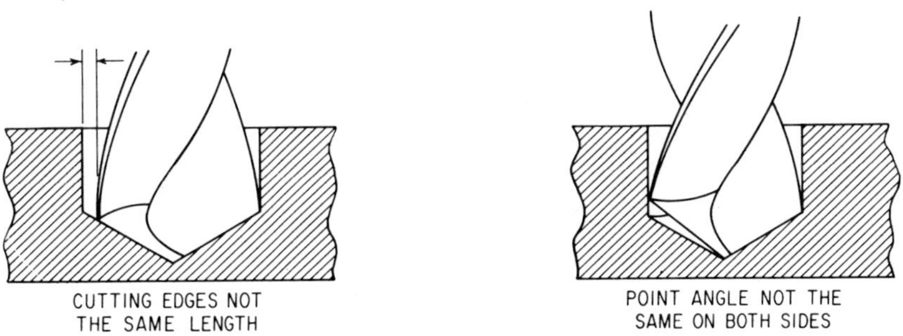

Figure 188 Effects of improper angles or centering

PART SIX

APPENDIXES

A

Printed-Circuit Information

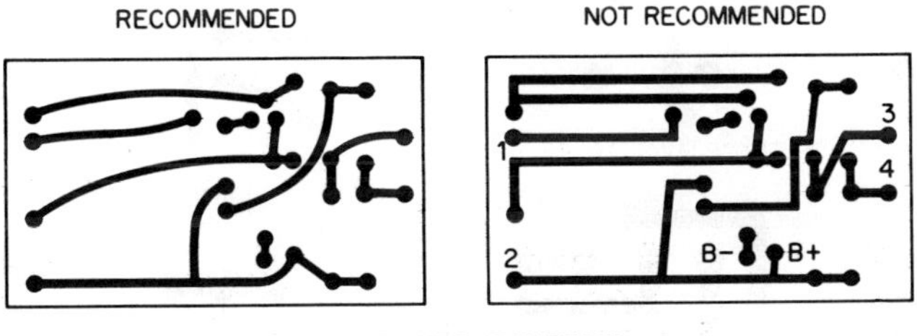

Figure 189 Printed-Circuit-Pattern Recommendations

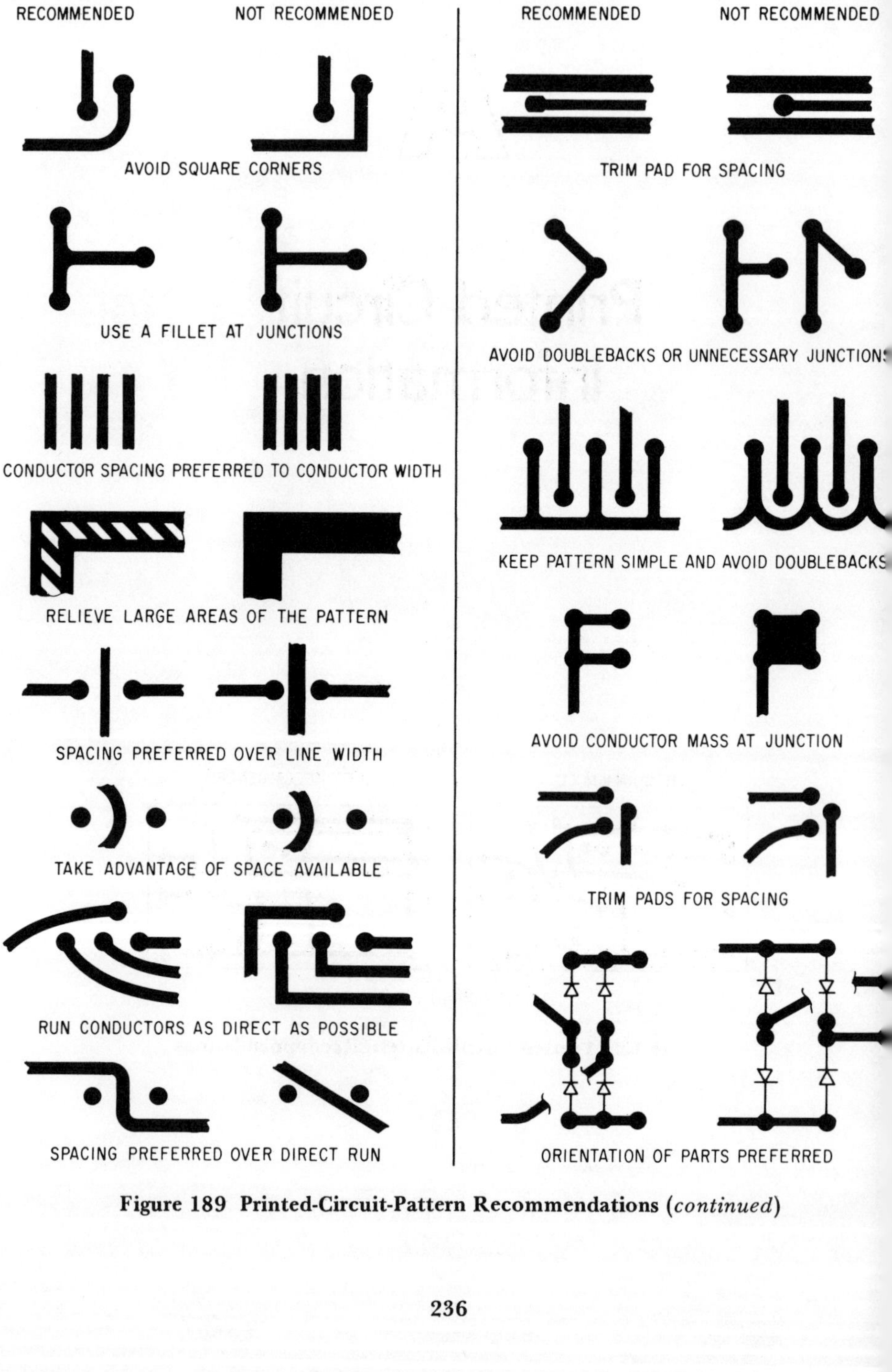

Figure 189 Printed-Circuit-Pattern Recommendations (*continued*)

Table A–1 COPPER FOIL WEIGHT

Thickness (in.)	*Weight of Copper Foil (oz) per Square Foot*
0.0007	0.5
0.0014	1.0
0.0027	2.0
0.0040	3.0
0.0068	5.0

Table A–2 CONDUCTOR WIDTH AND CURRENT RATING (for 1-oz copper)[a]

Conductor Width (in.)	*Approximate Maximum Current (A)*
0.020	0.14
0.040	0.30
0.060	0.44
0.080	0.60
0.100	0.85
0.125	1.10
0.150	1.30
0.200	1.80

[a] For 2-oz copper, multiply current values by 2.

Table A–3 DC RESISTANCE AND VOLTAGE DROP OF COPPER TRACES (20°C for 1-oz copper)[a,b]

Width of Traces	*Ohms per Inch*	*Ohms per Foot*	*Voltage Drop per 6″ of Trace*	
			mV at 100 mA	*mV at 3A*
0.01	0.049	0.58	29	—
0.015	0.032	0.39	19	—
0.020	0.024	0.29	15	—
0.050	0.010	0.117	6	180
0.1	0.005	0.058	3	90
0.25	0.002	0.023	1	33
0.50	0.001	0.012	.6	17

[a] For 2-oz copper, divide all values by 2.
[b] Resistivity of copper at 20°C is 10.4 Ω per circular mil foot or 8.168 Ω per square mil foot.

Table A-4 CONDUCTOR SPACING FOR PRINTED-CIRCUIT BOARDS

Peak Voltage Between Conductors (AC or DC)	*Minimum Spacing*[a] *(in.)*
0-150	0.025
150-300	0.050
300-500	0.100
500-1000	0.200

[a]Design specifications for adequate safety factor at sea level.

USE OF FERRIC CHLORIDE COPPER ETCHANT

The basic copper etching reaction using ferric chloride can be described as follows:

$$Cu + 2Fe^{3+} \rightarrow Cu^{2+} + 2Fe^{3+}$$

Essentially, this shows that the *copper* (metal) reacts with the *ferric ion,* which is in solution. This reaction yields a copper ion (cupric ion), which is also *soluble in solution.* From the above it can be seen that it is necessary for the ferric ion to be present in the solution before a reaction can take place. This ion is formed by the addition of water to the ferric chloride crystals. At times a small amount of HCL is added to the ferric chloride solution to help form this ion. Too much acid, however, will slow the copper etching action as a result of other factors that are involved. It is recommended by the author that only ferric chloride crystals and water be used.

Periodically, it is necessary to remove the sludge (Cu + ferric ion) that forms in the bottom of the etching vat. Merely pour off the top of the solution, clean out the sludge, and return the solution to the vat. A check should then be made on the density (Baumé reading) of the solution before it is used again. For the method used in making this check refer to the information sheet *Measurement of Solution Density*. Water or ferric chloride crystals should then be added to correct the density (solution content), depending upon the results of the test.

To increase the etching activity, the ferric chloride solution can be heated. The temperature is not critical and can be 70 to 130°F. It is important, because of side effects, not to heat the solution above the indicated temperature. Also, for consistent etching times, the solution temperature should be kept constant.

Adequate *solution agitation* should be used during the etching process. This is done to keep "fresh" solution flowing over the areas where the etching is taking place. Any sort of agitation can be used except that it must be remembered that the mechanical parts of an agitator can be attacked by the ferric chloride. This, of course, would destroy the agitator and also contaminate the solution. The simplest arrangement for agitation involves the use of air bubbles from a low-pressure air line. The air must be clean, free of oil and dirt, and is fed into the bottom of the vat through a perforated piece of plastic pipe. This method is illustrated in Figure 190.

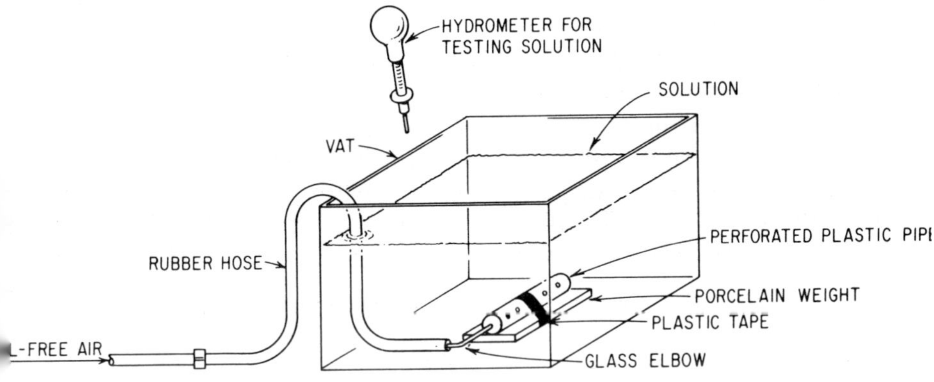

Figure 190 Air-agitation apparatus

CAUTION: For copper etching avoid the use of perchloric acid (HC_2O_4), hydrofloric acid (HF), or other nitric acid combinations. Unless they are handled properly, these chemicals are dangerous.

MEASUREMENT OF SOLUTION DENSITY (DEGREES BAUMÉ)

It is advantageous and reassuring to be able to make tests on a solution and establish a consistency in repetitious operations. Using *ferric chloride,* consistent etching times can be achieved by controlling the temperature and the *density or concentration* of the solution. The temperature is measured in degrees (Fahrenheit or Celsius) and the density is measured in degrees Baumé (written °Bé).

In using ferric chloride the measurement of the density is a check on the correctness of the solution. The ferric ion is a necessary part of the reaction with copper. Its presence in the solution is controlled by adding *water* or *ferric chloride* crystals. The solution cannot be too concentrated or too dilute. It must be nearly correct in order for etching to take place properly. A correct ferric chloride solution should have a density of approximately 1.38 (40°Bé).

The density (degrees Baumé) of a solution is simply measured with a *Baumé hydrometer.* These hydrometers are made to measure solutions with densities heavier than water (water has a density of 1.0 at 17.5°C). As is indicated by Table A-5, a 40°Bé solution of ferric chloride has a density of approximately 1.38. This is at the top of the range of an ordinary automobile battery hydrometer. If the need arises, the battery hydrometer could be used in determining the approximate density. Using the Baumé hydrometer the density is read directly from the scale.

Note that in the table the density and Baumé readings are taken at 17.5°C. The relationship of the two will change with temperature changes. It is therefore suggested that the density be measured before the solution is heated. After the density is corrected, the ferric chloride solution can then be heated up to 130°F. The Baumé

Table A-5 FERRIC CHLORIDE DENSITY AT 17.5°C

Density	*Degrees Baumé*	*Density*	*Degrees Baumé*	
1.00	0.00	1.21	25.16	
1.01	1.44	1.22	26.15	
1.02	2.84	1.23	27.11	
1.03	4.22	1.24	28.06	
1.04	5.58	1.25	29.00	
1.05	6.91	1.26	29.92	
1.06	8.21	1.27	30.83	
1.07	9.49	1.28	31.72	
1.08	10.74	1.29	32.60	
1.09	11.97	1.30	33.46	
1.10	13.18	1.31	34.31	
1.11	14.37	1.32	35.15	
1.12	15.54	1.33	35.98	
1.13	16.68	1.34	36.79	
1.14	17.81	1.35	37.59	
1.15	18.91	1.36	38.38	Nominal operating range (1.33–1.39)
1.16	20.00	1.37	39.6	
1.17	21.07	1.38	39.93	
1.18	22.12	1.39	40.68	
1.19	23.15	1.40	41.43	
1.20	24.17			

reading will drop about 0.1° for each 3°C change of temperature increase (refer to Figure 190).

DOUBLE-SIDED-BOARD REGISTRATION SYSTEMS

In the development of interrelated conductor patterns on a double-sided printed-circuit board it is necessary that these patterns be perfectly matched both in the *artwork development* stage and in the *fabrication* stage. This matching cannot be accomplished by guesswork, and it is necessary that a fixed registration system be used. In the development of the artwork and the exposure of the pattern on the board prior to etching, the alignment of the patterns on both sides of the board is critical in order to minimize feed-through problems and accomplish accuracy in drilling.

To obtain negatives for a double-sided board that match together, the pencil layout drawing of the printed-circuit pattern can be made up showing all the conductor runs for both sides of the board. For the conductor runs of one side use solid lines, and for the conductor runs of the other side of the board use dotted lines.

In taping the photomasters for both sides of the board, use a separate sheet of plastic or vellum for each side. Tape up the solid lines on one sheet and the dotted lines on the other sheet. When this is done, the pads common to both sides of the board will automatically be in the correct position. Use registration marks on the edge of each pattern for alignment of the patterns. Refer to Figure 191 for information on exposing the coated board with proper registration of the negatives.

NOTE: If the plate-through or the silk-screen process is followed, film positives rather than film negatives are used, and the process varies slightly but the same registration procedures can be used. Refer to the information sheets on the plate-through or silk-screen process.

Another very excellent system that can be used for obtaining coincident patterns is one that uses *black-white tape.* This tape is white on one side and black on the other. In the photographic process of making a negative using a high-contrast film, the white side of the tape will not show up as part of the pattern. This allows the circuit pertaining to one side of the board to be distinguished from the circuit on the opposite side. The areas of the circuit pattern

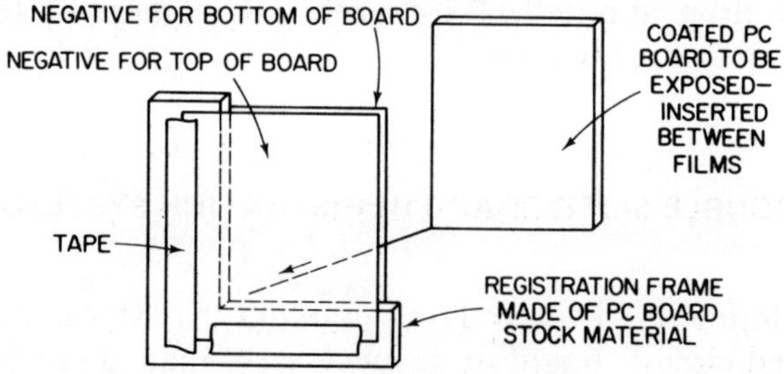

Figure 191 Film registration for double-sided exposures

that are common to both sides of the board are taped with a tape that is black on both sides. With this system the patterns can be made to match each other exactly if the films are held in registration during the exposure of the coated board. This system is briefly outlined in the following steps:

1. Obtain or develop a layout drawing of the circuit, showing the pattern for both sides of the board.
2. With a clear sheet of plastic or polyester film as an overlay, tape all the pads or circuits common to both sides of the board. Use tape and pads that are black on both sides. Off to the edge of the circuit, tape on some registration marks with black tape.
3. Using a black-white tape, tape up the balance of the circuit that is common to one side of the board. Let the white side of the tape face down and show through the back side of the plastic.
4. Turn the plastic or polyester film and the layout drawing over and tape the balance of the circuit on the opposite side of the board. Use a black-white tape with the white side facing the plastic.
5. Make or obtain photographic film negatives of each side of the taped-up plastic sheet.
6. Prepare both sides of the board with a light-sensitive emulsion such as KPR. Dry the board.
7. Tape the two film negatives together at one edge, being sure that the registration marks perfectly coincide. In some cases a strip of printed-circuit board can be taped in between the films at the edge.
8. Set up the board in an exposure frame with a film negative on each side. Clamp the film and board together with glass or plastic on each side.

9. Expose both sides of the board at once if that is possible. If not, see step 10.
10. Make a registration frame with the films taped to both sides (refer to Figure 191). With the films properly registered, the coated board can slip in between the films and be exposed one side at a time.
11. Continue subsequent operations.

PLATE-THROUGH CONSTRUCTION OUTLINE

On double-sided board (copper foil on each side of the laminate) the feedthrough connection from one side of the board to the other is made by use of eyelets or by some method of electroplating. This outline describes one of the plating processes, which uses a plated metal for the formation of the feed-through connection and also for the etching stopoff while etching the copper foil.

1. Drill or punch feed-through holes.
2. Clean the board.
3. Sensitize. Inside wall of holes and board are made conductive by deposition of copper by chemical reduction or *electroless plating*.
4. Apply negative resist pattern over background areas to act as plating stopoff. Carefully register patterns on both sides.
5. Plate with tin/nickel alloy.
6. Remove negative resist.
7. Etch the copper background.
8. Electrocoat with gold:
    a. To prevent edges of copper (at the side of the tin/nickel coat) from oxidizing.
    b. To cover contact fingers.
9. Cut board to size.
10. Mount parts.
11. Solder.
    a. Mask contact finger board and identification number.
    b. Dip at 450°F for 4 seconds.

12. Inspect and correct spots of bad soldering.
13. Wash or clean the board.
14. Spray on protective coating.

LAYOUT DEVELOPMENT CHECK LIST

The following check list is given so that the design of a printed-circuit layout can be evaluated. This is a fairly comprehensive list of items that need to be considered.

1. Is the conductor pattern orderly and uncluttered?
2. Are there conductor doublebacks?
3. Are the components oriented to be facing in the same direction wherever possible?
4. Are the components crowded?
5. Have large pattern areas been relieved?
6. Is there proper spacing between conductors?
7. Are the conductors run as short and direct as possible?
8. Is the conductor width correct?
9. Is the pad size correct?
10. Have jumpers been eliminated as much as possible?
11. Have fillets been used at junction points?
12. Have unnecessary or sharp corners been eliminated?
13. Does the photomaster agree with the schematic?
14. Is the polarity of the components indicated?
15. Is the size of the board correct or has the minimum board size been figured?
16. Are the component mounting holes (pads) located on grid intersection lines?
17. Is the distance between component mounting holes correct?
18. Have index holes been used properly?
19. Has the assembly of the components been considered?
20. Are terminals adequately spaced and properly located?
21. Has shielding been included?
22. Do the pads contain the proper-size component mounting holes?
23. Are adjustable components accessible?
24. Is there adequate spacing between the conductor pattern and mounting screws or mechanical parts?
25. Has the use of mechanical parts as part of the conductor pattern been avoided?

26. Has the electrical interference between conductors been considered?
27. Are heat-sensitive parts properly located?
28. Has feedback in high-gain circuits been considered?
29. Does the photomaster contain processing marks or dimension targets?
30. Are pattern registration marks used?
31. Is there clearance between the edge of the board and the pattern?
32. Is the board detail clearly indicated?
33. Is the photomaster viewed from the conductor side of the board?
34. Are tube sockets properly oriented?
35. Are transistors properly oriented?
36. Has servicing of parts been considered?
37. Have high-frequency characteristics been considered?
38. How is the board to be mechanically mounted?
39. Has the mounting of heavy parts been considered?
40. Is there physical interference between parts?
41. Have shock or vibration problems been considered?
42. Have environmental or humidity problems been considered?
43. Has lettering in the pattern area been avoided?
44. Are symbols for proper parts orientation used?
45. Are multilead components indexed?
46. Have provisions been made for labeling parts on the component sides of the board?

GLOSSARY FOR PRINTED CIRCUITS

Adhesion The bond strength of the conducting material (copper foil) to the substrate. Also refers to the bonding strength of the plies of laminated plastics.

Arc Resistance The number of seconds for a standard arc to establish a conductive path.

Art Master *See* Photomaster.

Back Side of Board The side of the board on which the copper foil is bonded. The conductor side of the board.

Base Refers to the sheet material impregnated with resin to form laminated plastics. This material can be cotton, paper, fiberglass, asbestos, or nylon fabric.

Canvas A heavy cotton fabric used in laminates.

Chemically Deposited Printed Circuit A printed-circuit pattern formed by a chemical reaction without an electric current.

Chemically Produced Printed Circuit A printed-circuit pattern formed by a chemical reduction.

Component Lead The lead attached to the component making an electrical connection to that component.

Component-Mounting Hole The hole drilled through the copper foil and substrate, in a pad or land, for the purpose of feeding through a component lead.

Component Side of Board The side of the board opposite to the conductor side of the board. The front side of the board.

Conductive Pattern A pattern of conductive material attached to an insulating substrate.

Conductor A conductive material in the form of a wire, plate, or strip. In printed circuit work a thin pattern of copper bonded to a laminate base.

Conductor Side of Board The side of the board on which the copper foil is bonded; the side opposite the component side of the board; the back side of the board.

Contact Fingers The portions of the conductor pattern that make contact in a plug-in receptacle. (*See also* Printed Contact *and* Tab.)

Copper Foil The thin copper sheet bonded to an insulating substrate.

Crosshatch The relief of large areas of the conductor pattern. This is done to prevent blistering of the copper foil.

Cure The process by which the physical properties of a material are changed by the application of heat or pressure.

Current Capacity The maximum electric current that can be handled by a conductor without undue heat, which would cause damage to the conductor or adjacent insulation.

Datum Points *See* Registration *and* Registration Marks.

Definition The sharpness of the edges of the conductor pattern. The sharpness of the detail with respect to the original artwork.

Detail Drawing The drawing that contains all necessary construction information, such as board specification, dimensions, size and location of holes, conductor pattern, and placement of the components.

Dimensional Stability Refers to the ability of a plastic part to retain its exact size or shape during or after molding or under conditions of temperature change.

Dimension Targets Accurately scaled marks, or the edges of marks, placed at the side of a photomaster or drawing for the purpose of representing a specific dimension. In the photoreduction process the photographer measures between these marks in setting up the required final size of the pattern.

Dip Soldering A method of soldering wherein the printed circuit, with mounted components, is quickly immersed into the surface of molten solder for the purpose of simultaneously soldering all component leads to the conductor pattern.

Dye A coloring used to make defects in an emulsion film easier to detect.

Dye-Veloper A commercial developer containing a dye that causes the undissolved emulsion to be colored.

Electrocoating The coating of an insulation surface, which allows the surface to act as an electrode for the purpose of electroplating. (*See also* Electrodeposition *and* Plating.)

Electrodeposition The deposition of a metal, by use of an electric current, from solution onto a metallic surface or an insulation surface that has been treated. This deposit forms a conducting path or a plating.

Embossed Foil Printed Circuit A type of printed circuit constructed by pressing a foil into depressions in the plastic substrate. The depressions follow the pattern of the circuit. The portion of foil that is not pressed into the depressed pattern is removed by mechanical means.

Epoxy Resin A synthetic thermosetting resin. These resins form a straight-chain molecular structure, based on ethylene oxide, by the condensation of bisphenol and epichlorohydrin.

Etchant A solution which, by chemical reaction, has the ability to dissolve or consume a solid material, usually metal.

Etched Printed-Circuit Board A type of printed circuit formed by chemically or electrolytically removing unwanted portions of copper foil, which is bonded to the substrate. A resist material is used as a stopoff for the part of the foil that is to form the conductor pattern.

Etching A chemical or electrolytic process whereby metal or material is removed in designated areas on an original piece.

Eyelet A hollow tube placed in a hole drilled through the substrate and pad and expanded on both ends. Used for the purpose of making an electrical connection from one side of the board to the other and also for holding the pad from lifting away from the board.

Feed-Through An electrical connection from one side of a printed circuit board to the other.

Filler Material The sheet material that is impregnated with a resin to form a laminated plastic. (*See also* Base.)

Fillet A rounded filling at a square corner.

Flame Resistance The ability to withstand high temperatures without supporting combustion.

Flexural Strength The ability of a material to withstand bending.

Flush Printed Circuit A type of printed circuit board in which the conductors of the circuit pattern are embedded into the base material so that the outer surfaces are flush.

Flux A natural rosin containing a mild acid. Used in soldering.

Front Side of Board The side of the board opposite to the conductor pattern. The component side of the board.

Funnelet A funnel type of eyelet. (*See also* Eyelet.)

Grid System Refers to the use of the intersection of horizontal and vertical lines for the accurate location of hole centers.

High-Pressure Laminate A laminate that is cured at a pressure between 1000 and 2000 psi.

Impact Strength The ability of a material to withstand shock under controlled test conditions.

Index Holes The holes placed in a board or marks on a drawing sheet used for the purpose of accurately locating the board or drawing for subsequent fabrication or drawing operations.

Indexing Notch A notch placed in one edge of a board for the purpose of ensuring accurate, predetermined contact connections.

Insulation Resistance The ability of a material to oppose the flow of electric current.

Interface Connection The connection of the circuit patterns on opposite sides of the board.

Jumper An electrical connection made between parts of the conductor pattern at places where it is impossible to include this connection as part of the pattern. This jumper connection is usually made of solid wire feeding through to pads from the component side of the board.

Laminated Plastics (Laminate) Refers to the single piece of plastic material formed by bonding resin-impregnated layer material together under heat and pressure. (*See also* Base.)

Land An expanded (circular) portion of the conductor pattern used for the purpose of making an electrical connection of a component lead or feed-through to the conductor pattern. (*See also* Pad.)

Lengthwise The direction on sheet material that is parallel to machine marks or warp thread direction.

Light Polymerization The process by which a material is changed from one form to another. In a photosensitive film exposure, the part of the film that is exposed to the light is made insensitive to the developer. The unexposed portions can be dissolved, leaving the underneath surface uncoated and subject to the action of an etchant. (*See also* Polymerization.)

Linen A light cotton fabric used in making plastic laminates.

Low-Pressure Laminate A laminate that is cured at a pressure below 1000 psi.

Master Drawing A scaled drawing used for the purpose of obtaining a photographic positive or negative. The master drawing is usually made to a double scale, a quadruple scale, or to some scale as great as 50:1. In the photographic reduction process the line definition is increased.

Master Pattern An accurately constructed pattern meeting the requirements specified in the master drawing. For use in the production of that particular printed-circuit board.

Melamine A synthetic thermosetting resin. Produced by the reaction of melamine and formaldehyde.

Module Refers to the grid system spacing. The intersection of the grid lines.

Moisture Resistance The ability of a material to resist the absorption of water.

Mounting Holes Holes in the circuit board for the purpose of mounting hardware.

NEMA Grades Grades of laminated plastics that have been set up by the National Electrical Manufacturers Association.

NEMA Standards Standards of characteristics or properties set up by the National Electrical Manufacturers Association.

Pad The word *pad* is interchangeable with the word *land.* (*See also* Land.)

Phenolic Resin A synthetic thermosetting resin. Produced by the condensation of phenol with formaldehyde.

Photomaster An accurately scaled copy or drawing from which a photographic positive or negative is made. This copy or drawing is usually expanded in size. A reduction to actual size is made in the photographic process.

Photoreduction The technique of reducing the size of a drawing. The final size of the drawing appears on the film in the photographic process.

Photosensitive Sensitive to light.

Photosensitive Emulsion A special solution made to be sensitive to light. When coated on a material and dried it forms a light-sensitive film (*See also* Light Polymerization.)

Pigtail Same as component lead.

Pinholes Small holes or defects in the conductor pattern.

Plated Printed Circuit A type of printed circuit in which the circuit pattern is plated onto a base material.

Plated-Through Hole An interface connection that has been formed by the electrodeposition of metal on the sides of a hole drilled through the base material between the conductor patterns.

Plating The method in which a metal is coated over another metal or material by the electrodeposition process.

Polymerization The change of a compound to another compound having the same elements but a higher molecular weight and different physical properties. This is done by the union of two or more molecules of the same kind.

Printed Circuit A unit made up of components, a conductor pattern, wiring, and certain hardware attached to an insulating base material.

Printed-Circuit Board An insulating base material with a thin copper foil bonded to it. The copper foil is etched out according to a pattern forming the conductors for a circuit.

Printed-Circuit Component A circuit component especially designed to fit onto a printed-circuit board.

Printed Contact Refers to that portion of a printed-circuit pattern that makes connection with a spring-loaded contact in a plug-in type receptacle. (*See also* Contact Fingers.)

Registration A system of guide pins or marks whereby two or more patterns can be accurately located or positioned with respect to each other. In printed-circuit work the circuit pattern for one side of the board is positioned with respect to the pattern for the opposite side of the board.

Registration Marks The marks on the photomaster drawing, the film negative, the film positive, or other artwork that are used for the purpose of positioning various copies of these drawings with respect to each other.

Resin A natural or synthetic, solid or semisolid organic product with a high molecular weight and no definite melting point. In the formation of laminates the resin is impregnated into the filler or base material and cured under heat and pressure. The chief resins are the phenolics, melamines, epoxies, and a silicone resin.

Resinoid Refers to any of the synthetic resins, whether they are in the uncured state or in the cured state.

Resist An ink, paint, or a plated material that is used as a stopoff, which protects the desired portions of printed circuit from the action of the etching solution.

Resolution The sharpness and clarity of the detail of a reproduction.

Shielding Conductive areas used to isolate or eliminate interference between parts of the circuit.

Silicone A polymeric material in which oxygen and silicone atoms form the main chain of the chemical makeup. Produced from silica (sand) and methyl chloride.

Solderability The ability of a material to accept solder in a wetting action.

Sprayed Printed Circuit A type of circuit in which the circuit pattern is formed by spraying molten conductor material.

Stamped Printed Circuit A type of circuit formed by die stamping the foil to obtain the circuit pattern.

Stopoff A material or substance used for the purpose of preventing chemical action in certain areas. Refers to plating stopoff or etching stopoff.

Substrate The insulating laminated plastic piece to which the copper foil is bonded.

Surface Resistance The ability of the material to oppose the flow of electric current across or over the surface of the material.

Synthetic Resin A complex, organic, semisolid material built up by the chemical reaction of relatively simple compounds.

Tab An extending part of a printed-circuit board containing printed contacts, which match with spring contacts in a plug-in strip.

Terminal A ridged metallic post connected to the circuit pattern, which is used for the purpose of connecting external leads or jumper wires to the pattern. The terminal is mounted on the component side of the board, fed through, and soldered to the pattern on the conductor side of the board.

Thermal Conductivity The ability of a material to conduct heat.

Thermosetting Plastic A plastic that hardens when heated but does not soften when reheated.

Undercut Refers to the action of the etchant under the edge of the resist causing the pattern size to vary. This is considered to be an undesirable effect.

Vapor-Deposited Printed Circuit A process in which a printed circuit is formed by the condensation of material from a gaseous state.

Water Absorption The ratio of the weight of water absorbed by a material to the weight of the dry material.

B

Supplementary Information

MISCELLANEOUS TYPES OF CHASSIS

In addition to the basic types of chassis covered in preceding assignments, many other types can be used. The type of chassis usually will be determined by the *kind of tools available,* the *skill* or experience of the technician, or by the *circuit* itself. Many factors need to be considered.

Variations can be incorporated in each of these chassis types. This is often necessary in order to adjust the basic plan to fit individual situations.

Several examples of miscellaneous chassis types are given in Figures 192 to 196.

Purpose of Assignment

1. To gain information on various basic types of chassis that can be constructed in the ordinary school chassis-construction shop.

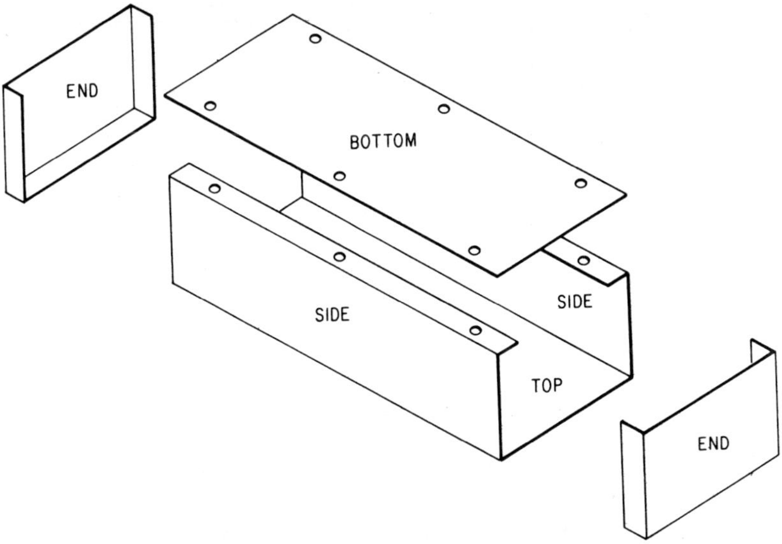

Figure 192 Channel type of chassis

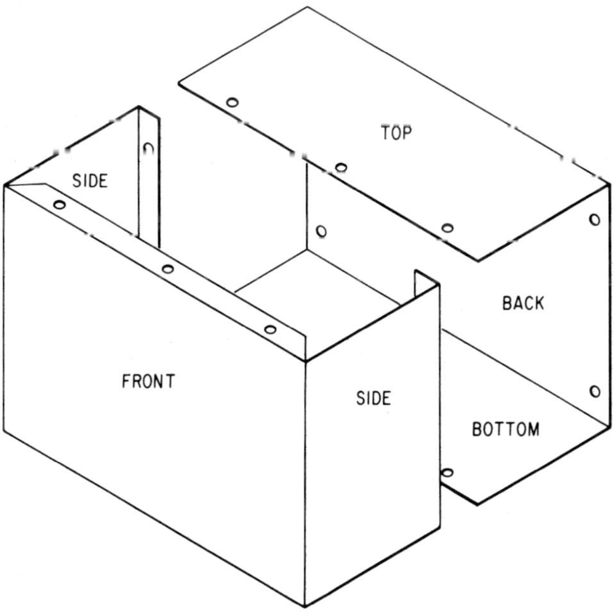

Figure 193 Double wraparound chassis

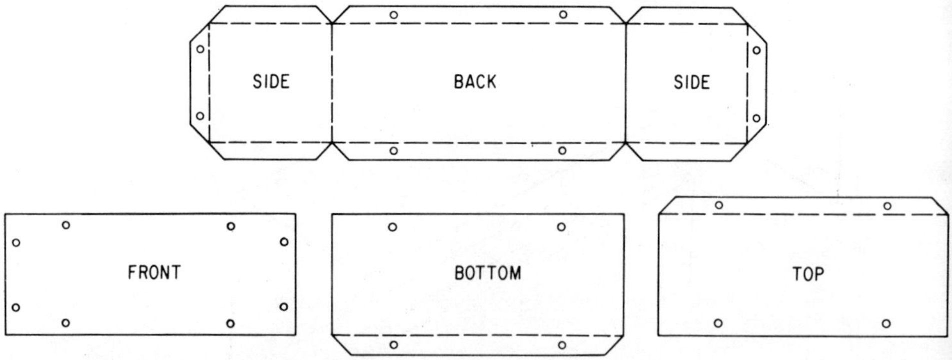

Figure 194 Wraparound side, back, and side

2. To gain experience in making an adaptation of a chassis to a particular circuit.
3. To gain experience in determining and making a complete set of chassis layout plans.

Tools and Equipment

Tools and equipment as required by individual choice of the chassis to be constructed.

Materials

Various materials determined by the choice of chassis to be constructed.

Assignment

Decide on the type of chassis that can be adapted to the requirements of a particular circuit. Refer to *Prototype Development Sequence* (Appendix B) for a listing of the factors to be considered.

Directions

1. Work out *all* the detail on the adaptations of the chassis to the chosen circuit.
2. Make complete layout drawings of the chassis. Determine all dimensions, the proper layout pattern, and any necessary construction information.
3. Construct the chassis. Be sure to think ahead and follow a proper sequence in completing the hand- and machine-tool operations that are involved.

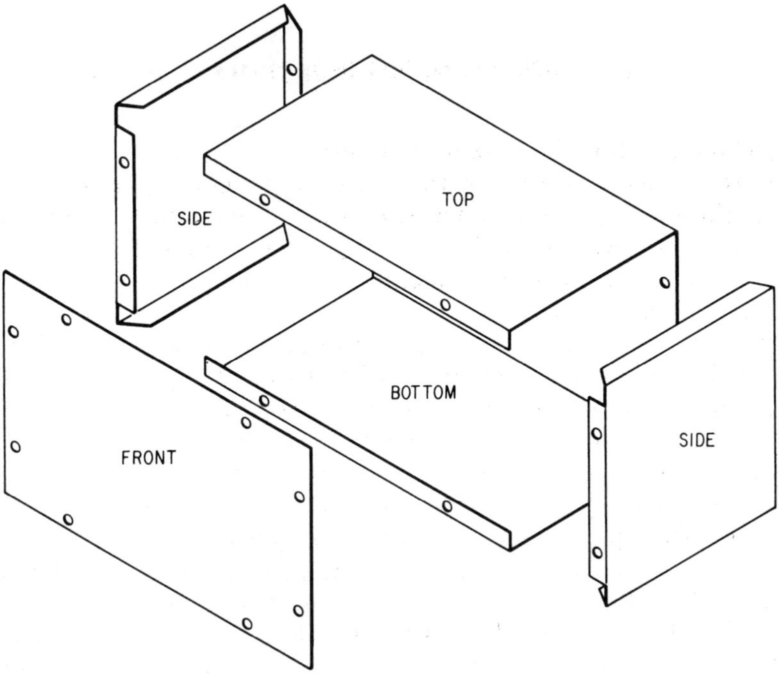

Figure 195 Wraparound top, back, and bottom

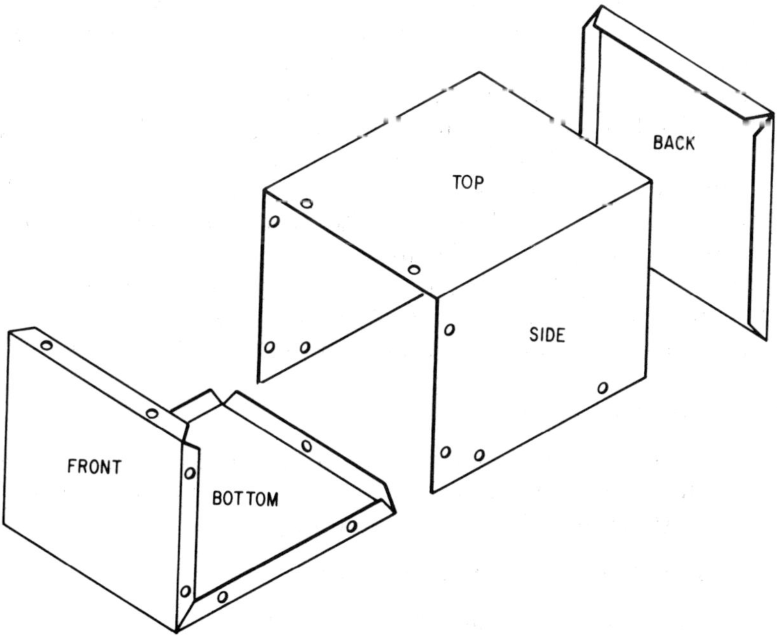

Figure 196 Front and bottom piece chassis

A technician should be skillful in using the tools or instruments that are involved with his work; therefore, he must learn something about those tools or instruments, become acquainted with the parts, and have a knowledge of their operating capabilities. This is true in the use of electronics instruments or mechanical devices such as machine or hand tools.

Part of a person's skill in the use of an instrument or tool is his knowledge of the *adjustments* associated with that particular device. This knowledge is also related to the *use* of the device and is quite important in order that it not be damaged in any way by its use. Some of this information can be figured out on a logical basis, or in many cases instruction manuals can be used as a reference.

It is often quite easy to visualize what must be done to properly care for mechanical tools or devices. Damage to parts can be easily seen most of the time, and parts that need to be lubricated can be located with very little effort. A wrench should be used for loosening and tightening bolts and the parts should be used *only* for their intended use. A regular system of lubrication should be set up in order to properly care for those machines that are in constant use.

Care of equipment in a shop includes *cleaning up* the equipment itself and also the immediate surrounding area. No person can work efficiently or to the best of his or her ability if the machine or the things that he or she is touching are covered with old chips or dirty grease and oil. Equipment must be kept clean, and the person who uses the equipment should clean up the area.

A description of the various tools used in most shops is given below. These tools are basic and the most commonly used tools. Other tools will no doubt be found in some shops. The care and general use of any tool that is used will be similar to that of the tools that have been shown here. A sense of responsibility and good judgment are essential to the proper care and use of any tool.

Wire and Sheet-Metal Gauge

The wire and sheet-metal gauge is shown in Figure 197. The essential features of this tool are shown and the important parts are identified. Make special note of the following:

Use of the Wire and Sheet-Metal Gauge

1. Measuring the gauge number of solid wire.
2. Measuring the gauge (thickness) of sheet metal.

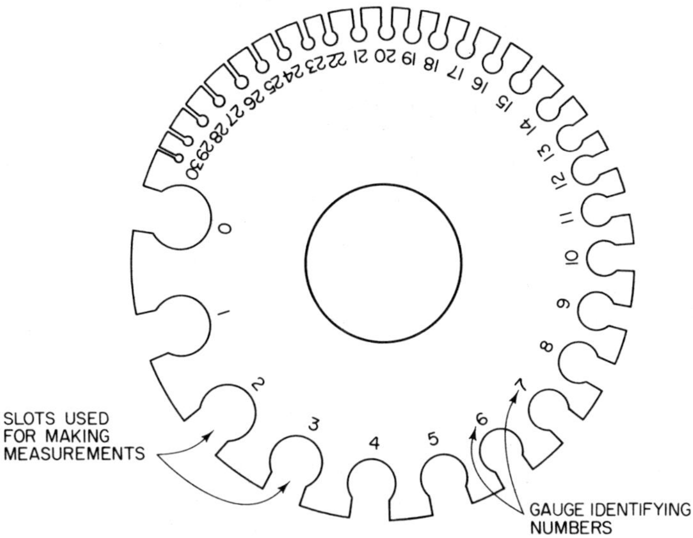

Figure 197 Brown and Sharpe wire gauge

Important Adjustments

None applicable.

Precautions

1. Do not force the wire or sheet metal into one of the slots.
2. Be sure that the tool is held at a right angle to the wire or sheet-metal piece.

Twist-Drill and Machine-Screw Gauge

The twist-drill and machine-screw gauge is shown in Figure 198. The essential features of this tool are shown and important parts are identified. Make special note of the following:

Use of the Twist-Drill and Machine-Screw Gauge

1. Identifying the size of twist drills.
2. Identifying the size of holes required.
3. Obtaining tap-drill and body-drill information.

Important Adjustments

None applicable.

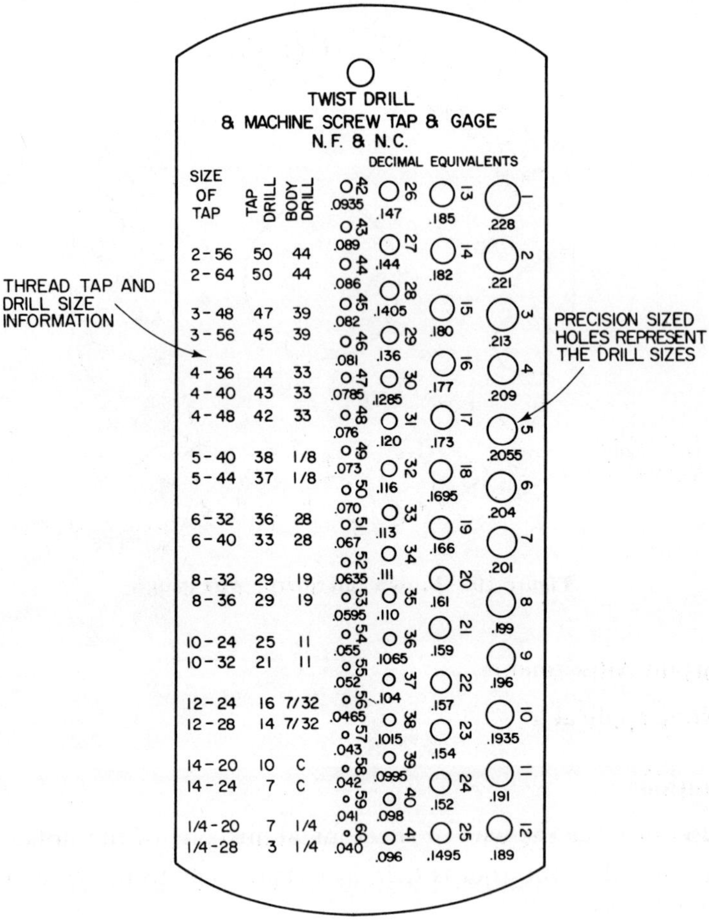

Figure 198 Twist drill and machine screw tap and body drill gauge

Precautions

1. Do not force drills into the holes.
2. Do not force the drills sideways or back and forth while the drills are in the holes.

Micrometer

The micrometer is shown in Figure 199. The essential features of this tool are shown and important parts are identified. Make special note of the following:

Use of the Micrometer

Making precision measurements of diameters or thicknesses.

Important Adjustments

1. Check to see if the zeros line up properly when the spindle is touching the anvil.
2. Check the position of the lock nut.
3. Check the tightness of the ratchet.

Precautions

1. Do not drop the tool.
2. Do not tighten the spindle down tight on to the workpiece.
3. Store the tool properly when it is not being used.

Turret Punch

The turret punch is shown in Figure 200. The essential features

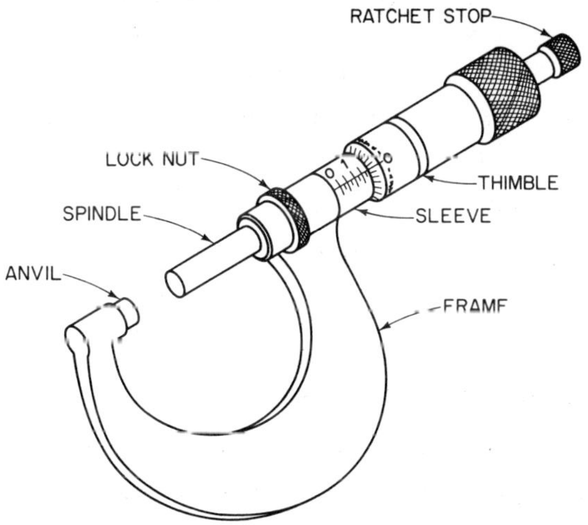

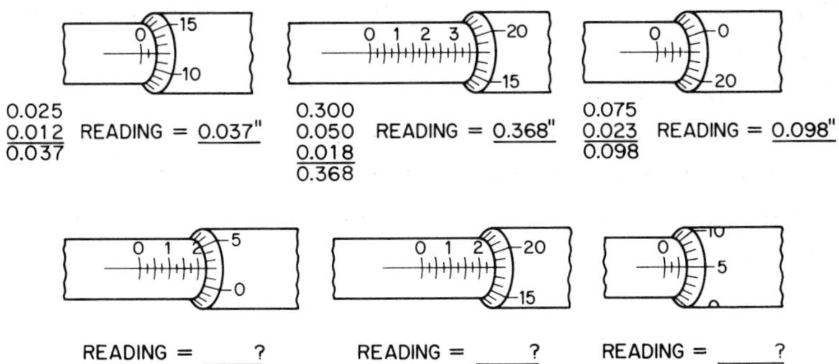

Figure 199 Examples of reading a micrometer

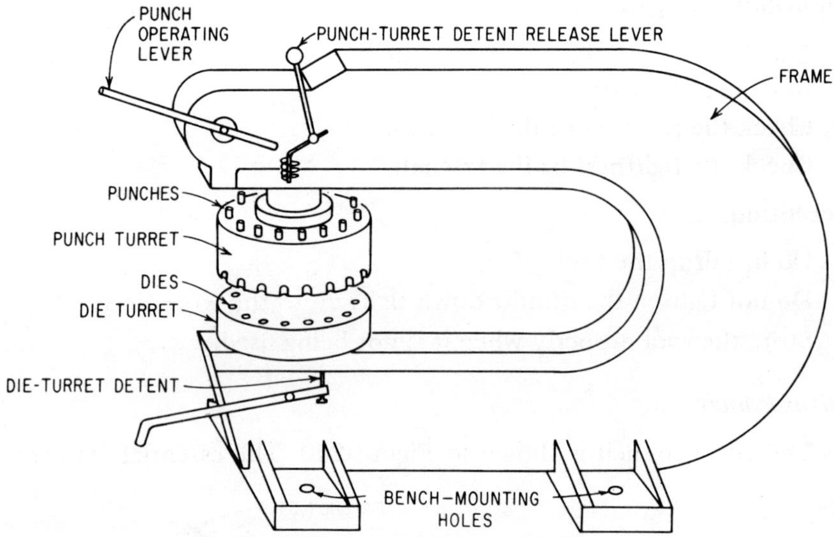

Figure 200 Turret punch

of this tool are shown and important parts are identified. Make special note of the following:

Use of the Turret Punch

1. Punching square or round holes.
2. Punching transformer and capacitor cutouts.
3. Notching.

Important Adjustments

1. If applicable, adjust the punch to match the die.
2. Select the proper punches and dies by rotating the turrets.

Precautions

1. Make sure that the punch exactly matches the die as they go together.
2. Use oil on the punches.
3. Construct a V-notched tool to be placed over the work when the punch is pulled out of the punched-out hole.
4. Be sure that the enter tip of the punch is located properly in the center punch mark before punching.

Squaring Shear

The squaring shear is shown in Figure 201. The essential features of this tool are shown and important parts are identified. Make special note of the following:

Use of the Squaring Shear

1. Cutting straight edges for reference or finish cut.
2. Squaring or trimming edges.

Important Adjustments

1. Adjust side guides to be exactly 90° to cutting blade.
2. Adjust hold-down bar pressure to be equal on both sides.
3. Adjust foot-operated treadle to pull the cutting blade down evenly.
4. Adjust relative position of cutting blade to the bed.

Precautions

1. Keep feet from under the foot-operated treadle.
2. Keep fingers away from cutting blade.

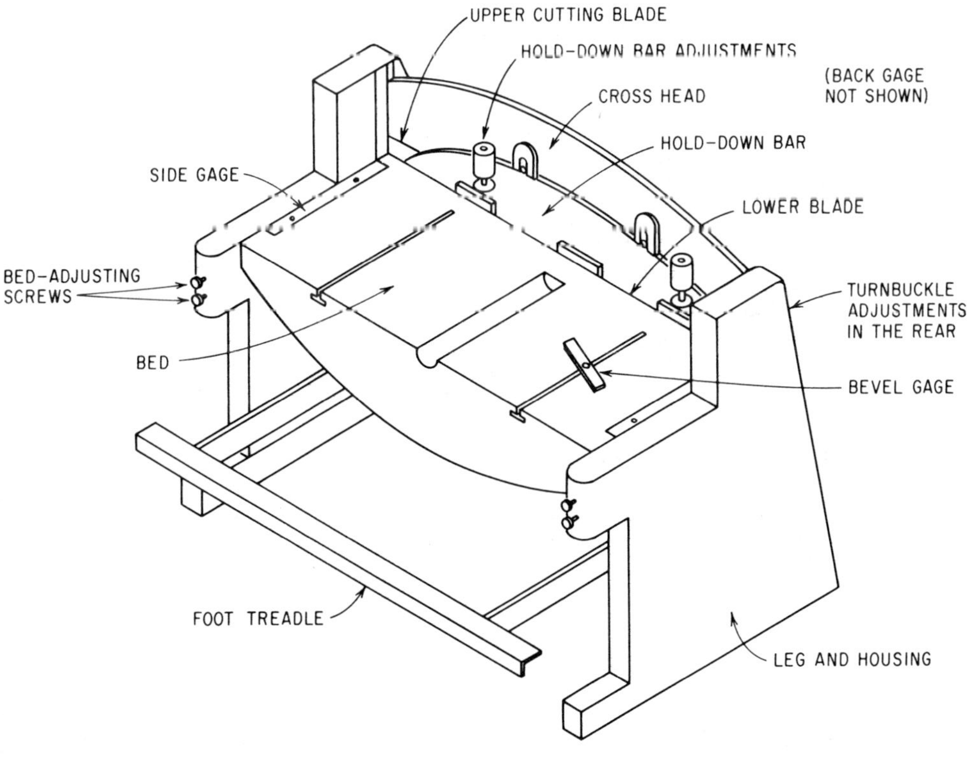

Figure 201 Foot-operated squaring shear

Press Brake

The press brake is shown in Figure 202. The essential features of this tool are shown and important parts are identified. Make special note of the following:

Use of the Press Brake

1. Bending the sides of fairly deep boxes.
2. Bending lips or tabs on the edges of panels.
3. Making offset bends.
4. Making rounded bends or sharp bends.

Important Adjustments

1. Adjust the die rack to be parallel to the punches.
2. Clamp the proper combination of punches tightly in place.

Precautions

1. Do not drop the punches when loosening the clamping bolts.
2. Use special care when setting up the punch to meet the bend line.

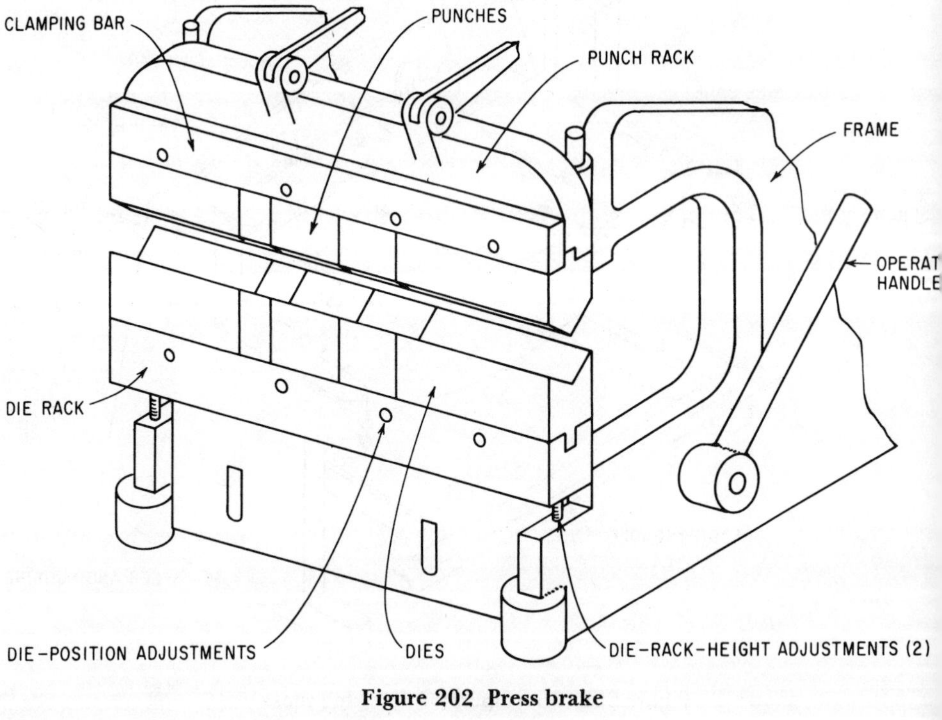

Figure 202 Press brake

Wing Brake

The wing brake (sometimes referred to as a *box-and-pan brake* or *finger brake*) is shown in Figure 203. The essential features of this tool are shown and important parts are identified. Make special note of the following:

Use of the Wing Brake

1. Bending the sides of shallow boxes.
2. Bending lips or tabs on the edges of panels.
3. Making an offset bend.

Important Adjustments

1. Adjustment of the hold-down pressure.
2. Clearance adjustment between the end of the "fingers" and the fold-up bar.

Precautions

1. Do not bend too heavy a gauge. Depending on the length, 20-gauge steel or 12-gauge aluminum can be handled.
2. Make adjustments *even* on both sides of the tool.

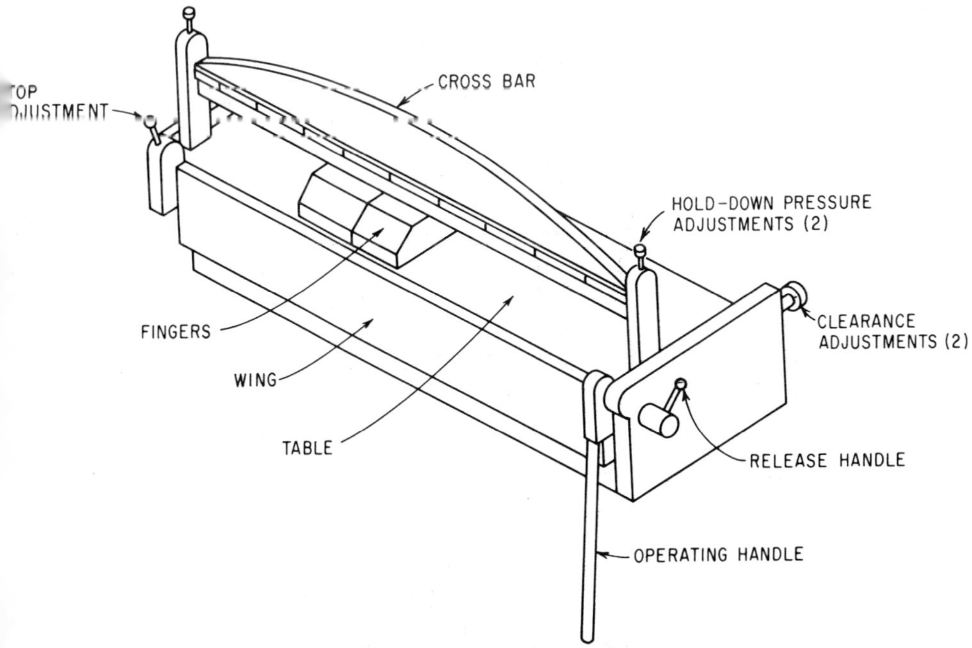

Figure 203 Wing or box-and-pan brake

Corner Notcher

The corner notcher is shown in Figure 204. The essential features of this tool are shown and important parts are identified. Make special note of the following:

Use of the Corner Notcher

1. Making a 90° corner notch prior to bending up the sides of a box.
2. Cutting a shearing notch or shearing a short length.
3. Cutting a 90° notch or a 45° trim cut.

Important Adjustments

1. Adjust the relative position of the cutting edges by using the lower die adjustments.
2. Adjust the tightness of the vertical block on its track.

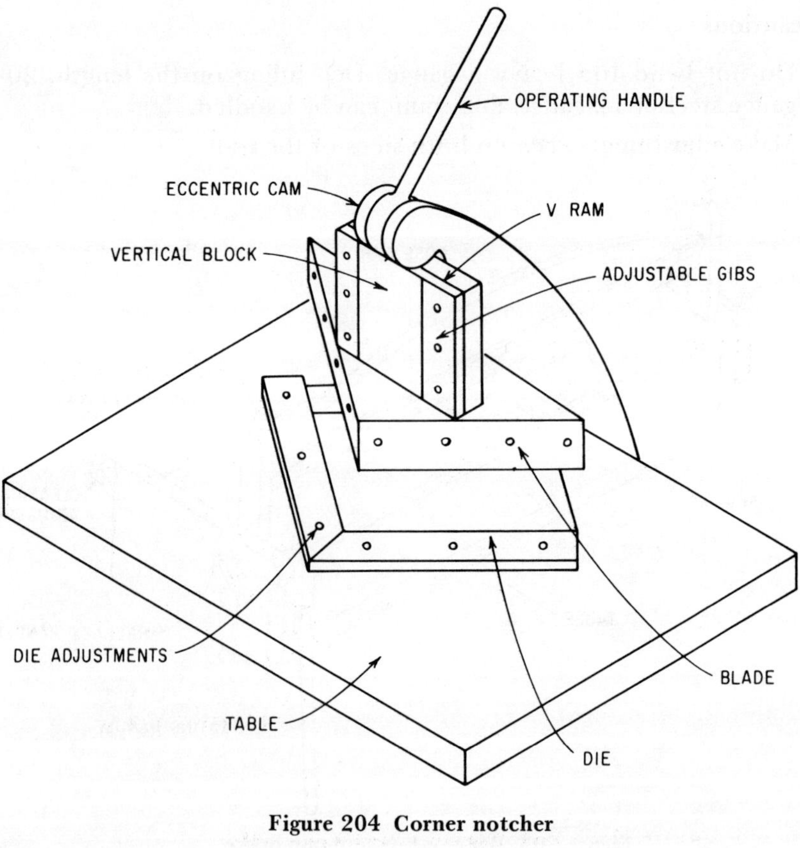

Figure 204 Corner notcher

Precautions

1. Be sure that the blade and die alignment is correct.
2. Keep fingers from under the cutting edges.

Standoff Depressor

The standoff depressor is shown in Figure 205. The essential features of this tool are shown and important parts are identified. Make special note of the following:

Use of the Standoff Depressor

1. Making standoff dimples.
2. Making depressions to act as chassis feet.

Important Adjustments

1. Be sure that the die matches the punch.
2. Use guidelines for locating the position of the depression.

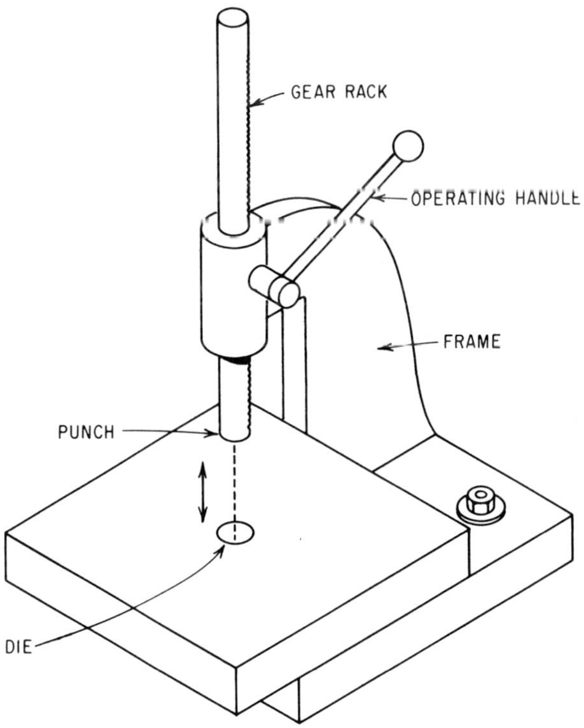

Figure 205 Standoff or dimple depressor

Belt Sander

The belt sander is shown in Figure 206. The essential features of this tool are shown and important parts are identified. Make special note of the following:

Use of the Belt Sander

1. Making the edges of a piece smooth and straight.
2. Trimming a piece to size.
3. Shaping a piece to meet requirements.

Important Adjustments

1. Adjust the belt for proper tightness.
2. Adjust the pulleys for correct alignment so that the belt will not creep sideways.
3. Adjust the worktable for the proper tilt.

Precautions

1. Do not sand soft or gummy material.
2. Hold the workpiece against the worktable while sanding.
3. Sand the workpiece slowly.
4. Be sure that your hands don't accidentally slip into the belt.

Louver Punch

The louver punch is shown in Figure 207. The essential features of this tool are shown and important parts are identified. Make special note of the following:

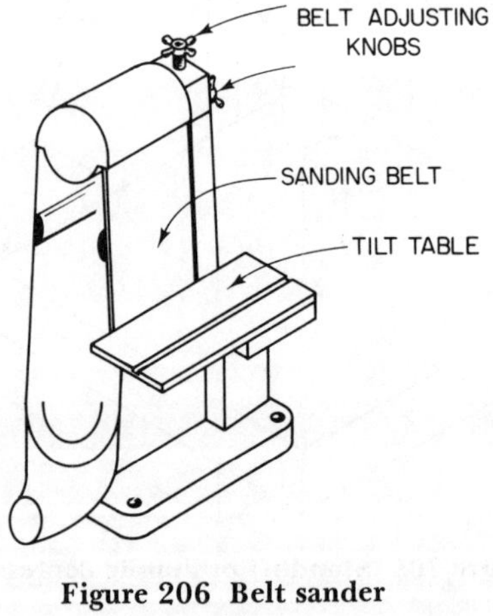

Figure 206 Belt sander

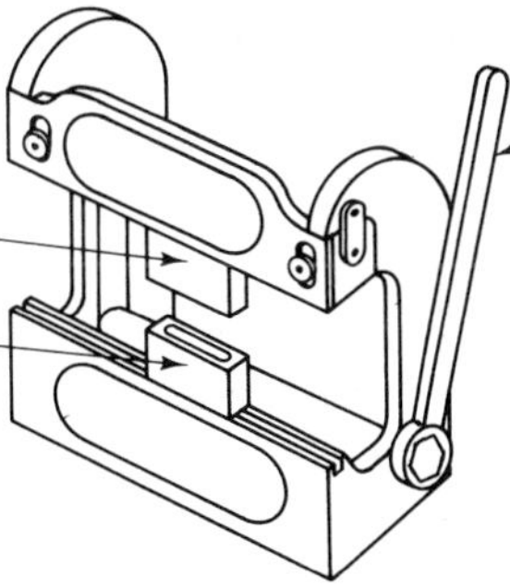

Figure 207 Louver punch

Use of the Louver Punch

Punching ventilation louvers in a chassis.

Important Adjustments

1. Be sure that the punch matches the position of the die.
2. Adjust the guides to be at the proper angle if they are used.
3. Adjust the depth of the punch by using shims as required.

Precautions

1. Make trial louvers on scrap material before punching the workpiece.
2. Pull, operating smoothly but firmly.

PROTOTYPE DEVELOPMENT SEQUENCE

The development of an electronics prototype unit must necessarily follow some sort of a plan. Working out the detail and thinking through the interrelationship of the factors that are involved will save costly, time-consuming errors.

In the development sequence, there should first be a comprehensive *mental evaluation* of the many factors involved, and several *decisions* should be made before written plans are begun. "Breadboard" the circuit to confirm preliminary proposals. *Detailed plans* should be drawn up before construction is started. Following the completion of the unit, it should be tested and given *operational checks.* A complete descriptive *report* should then be compiled and submitted with the unit.

Depending upon the circuit, there are many things that *must* be considered. Also, depending upon personal preferences, there are many things that *could* be considered. The possible detail involved in the development sequence proposed above is given in the following outline:

I. **Preliminary Decisions and Plans**
   A. Decide on what is to be made.
      1. Use or purpose of the unit.
         a. Reason for making it.
         b. Probability of success.
      2. Availability of parts.
      3. Cost.
   B. Decide on the type of construction.
      1. Printed circuit.
      2. Wired circuit.
      3. Combination.
   C. Decide on the type of chassis.
      1. From the aesthetic or human engineering standpoint.
         a. Style and color.
         b. Composition, design, or panel markings.
         c. Placement of controls, meter location, jack locations.
      2. Cabinet and panel exterior.
         a. Finish and texture.
         b. Method of marking panel.
         c. Parts-mounting facilities.
         d. Handles, footing, shock mounting.
         e. Speaker grill, heat ventilation.
      3. Component layout.
         a. Signal logic, space utilization.
         b. Electric and magnetic shielding, electric or magnetic interference.
         c. Weight distribution.
      4. Component mounting.
         a. Terminal boards, terminal strips, connection points for wires, standoffs.
         b. Interconnecting cables, types of connectors.
         c. Mounting brackets for large components, hardware, and vibration protection.
      5. Maintenance.
         a. Chassis or panel removal, accessibility, component removal, and fuse replacement.
         b. Operability when open, test points.
         c. Component labeling, coding of wires and terminals.

6. Safety
   a. Fuses, thermal protection, interlocks, isolation, pilot lights.
   b. High-voltage covers, feedthrough grommets, feedthrough insulators.
   c. Hardware technique, operation, lock washers.
   d. Marking dangerous points.

II. Breadboarding the Circuit

A. Determine approximate location of all components.
B. Make preliminary operational checks.
C. Determine the probability of success.
   1. Will the unit work properly?
   2. Can you build it properly?

III. Written Plans in Detail

A. Make layout drawings of chassis complete in *all* details.
B. Draw up wiring pictorials showing parts placement and wiring.
C. Determine routing and connection information for cabling, if applicable.
D. Make exploded views of assembly detail.
E. Draw various views of the proposed finished appearance.
   1. Isometric or oblique drawings.
   2. Front, back, or side views.

IV. Fabrication or Construction

A. Perform machine- or hand-tool operations.
   1. Layout, punching, drilling, forming, deburring, and fastener installation.
   2. Bracing for structural strength.
   3. Marking panel, painting, other finishes.
B. Mount major components.
C. Construct mechanical details.
   1. Dial cabling, pulleys, special controls.
   2. Special mechanical levers and devices.
D. Wire the chassis.
   1. Wire dress, cabling, breakout, color code, insulation setback, wire wrap, connection points.
   2. Twisting AC leads, shielding.
   3. Cable lacing.
E. Solder the connections.
   1. Flow, contour, wicking, thermal damage, neatness.
   2. Flux removal, cleanliness.

F. Complete final detail.
   1. Mechanical adjustments.
   2. Recheck all work.

V. **Operational Checks and Testing**
   A. Adjustments and mechanical tests.
      1. Calibration, alignment.
      2. Other adjustments.
   B. Run operational electrical tests.
      1. Operation ease.
      2. Distortion, output capabilities, stability, operation to specifications, sensitivity tests, operational limits, power consumption.
      3. Data collection, graphs, charts.

VI. **Written Report**
   A. Include description, specifications, and instructions.
      1. Title page.
      2. Table of contents.
      3. Proposed finished appearance.
      4. General description.
      5. Technical analysis.
      6. Schematic drawings.
      7. Mechanical drawings.
      8. Performance data.
      9. Operating instructions.
      10. Parts list.

SWAGE-NUT MOUNTING

Swage nuts are used for the purpose of providing strong load-bearing threads in thin metal and for blind fastening in inaccessible areas. They are easy to install, are often made of hardened material, and can be used in steel, alloys of copper, and aluminum.

There are many types of swage nuts. Figure 208 shows the cross section of a typical flush-mounting type and the essential features of its construction. In addition to this type, there is the self-locking type, a feedthrough type, and a type that requires the use of a tool for its mounting. The swage ring is often knurled or serrated to prevent the nut from turning when in its mounted position (refer to Figure 208).

Because of the structure of this type of nut and the method where-

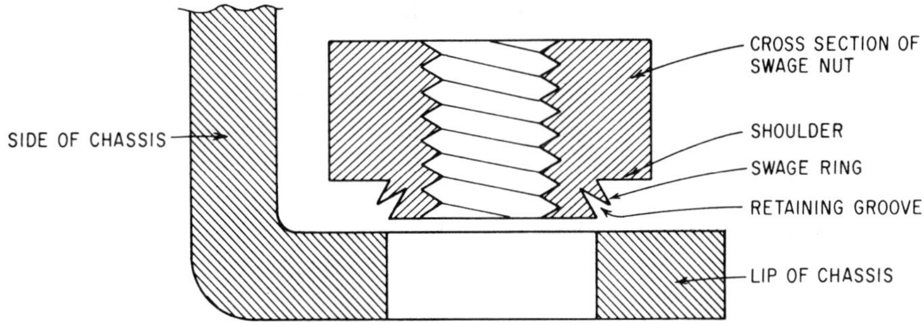

Figure 208 Cross section of typical swage nut

by it is held in place, the mounting hole must be carefully machined to a tolerance of ±0.002″. Information on the size of the hole must be obtained from the manufacturer's instruction sheets.

The following steps give a probable sequence that would be followed in the mounting of this type of nut in a chassis piece:

1. Make the layout of hole centers and cut lines on the blank stock.
2. Punch, drill, or ream various holes, make cutouts, and cut the outline of the piece.
3. Bend the lips upon which the swage nuts will be mounted.
4. Place the nut facing in the correct direction in the mounting hole.
5. Locate flat surfaces on the top and bottom of the nut as shown in Figure 209.
6. Apply pressure using a vise or press. Do not squeeze the nut too far into the metal but just far enough to cause the shoulder to meet the surface of the metal.
7. Bend up the sides of the chassis, if applicable.

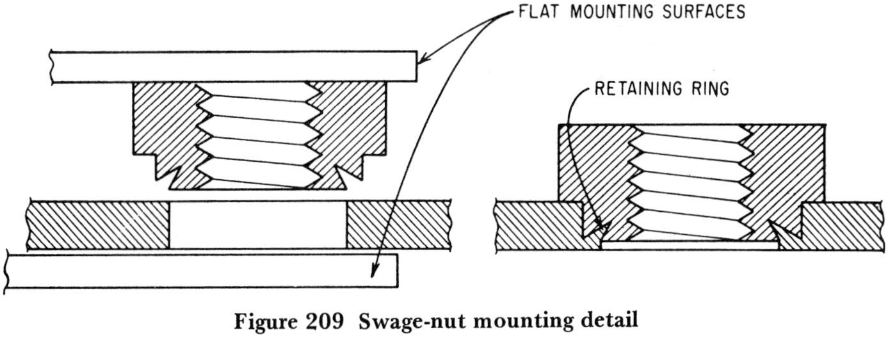

Figure 209 Swage-nut mounting detail

WIRE-WRAP

Wire-wrap techniques have been used for many years for making electrical wire connections to terminal blocks, printed-circuit boards, connector plates, and so on. When done properly, this type of connection is a high-quality electrical connection, even though it is solderless. Wire-wrap connections are fairly quick and easy to make. When used in conjunction with integrated-circuit chips, wire-wrap techniques can contribute to a very heavy wiring density, where many wire connections can be applied in a small space. Often, completely automatic machines are used on mass-production assembly lines.

In wire-wrap connections, a square terminal pin is used and wrapped with a 22- to 30-gauge solid wire (see Figure 210). Wrapping the wire around the square pin creates a considerable number of high-pressure contact points at the places where the wire bends around the square corners of the pin. These high-pressure points are what contribute to a low-resistance, high-quality electrical connection.

A proper connection will include five to six turns of bare wire followed by 1 to 1½ turns of wrap with the insulation still on the wire. The first wrap should be placed as close to the bottom of the pin as possible. This allows up to three or four wraps per pin.

For hand wire wrapping, an electrical or pneumatic wire-wrap gun (tool) is used for obtaining a proper wire-wrap connection on the terminal pin. The wrap of wire should be smooth and even, with no space or overlaps between the turns of wire.

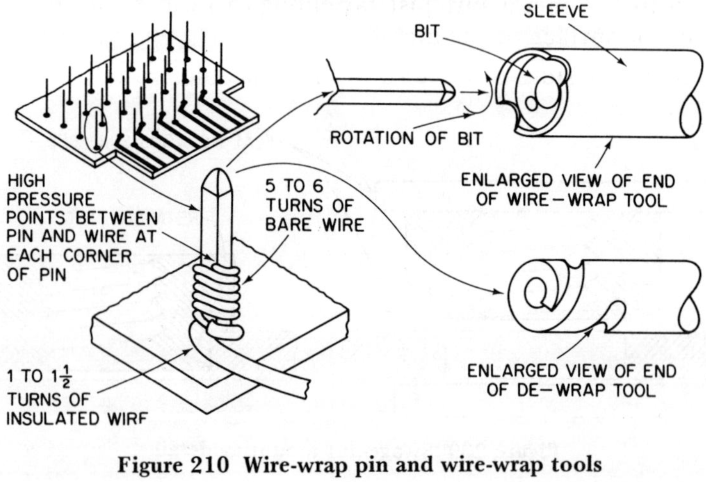

Figure 210 Wire-wrap pin and wire-wrap tools

The nose of the gun contains a bit that rotates inside a surrounding sleeve (see Figure 210). The wire is stripped approximately 1¾″ and inserted into the outermost hole of the bit, and the gun is then slipped over the pin with the pin entering the center hole. When the bit is rotated, the wire will then wrap tightly around the pin.

For removing a wire-wrap connection, a de-wrap tool, with a specially designed curved hook, is used to unwrap the wire from the pin. Figure 210 shows a view of the end of the de-wrap tool.

Various-sized wires may be used, and usually the wire is color-coded for identification purposes. Special wire strippers are used to prevent the wire from being nicked. Some tools premeasure the wire for the needed length on the end of the wire that is stripped.

C

Charts and Tables

Table C-1 CELSIUS-FAHRENHEIT TEMPERATURE CONVERSION[a]

°C	*Known Temperature*	°F	°C	*Known Temperature*	°F
-17.78	0	+32.0	12.21	54	+129.2
16.65	2	35.6	13.32	56	132.8
15.54	4	39.2	14.43	58	136.4
14.43	6	42.8	15.54	60	140.0
13.32	8	46.4	16.65	62	143.6
12.22	10	50.0	17.76	64	147.2
11.10	12	53.6	18.87	66	150.8
10.00	14	57.2	20.00	68	154.4
8.88	16	60.8	21.09	70	158.0
7.77	18	64.4	22.20	72	161.6
6.66	20	68.0	23.31	74	165.2
5.55	22	71.6	24.42	76	168.8
4.44	24	75.2	25.53	78	172.4
3.33	26	78.8	26.64	80	176.0
2.22	28	82.4	27.75	82	179.6
1.11	30	86.0	28.86	84	183.2
0.00	32	89.6	30.00	86	186.8
+1.11	34	93.2	31.08	88	190.4
2.22	36	96.8	32.19	90	194.0
3.33	38	100.4	33.30	92	197.6
4.44	40	104.0	34.41	94	201.2
5.55	42	107.6	35.52	96	204.8
6.66	44	111.2	36.63	98	208.4
7.77	46	114.8	37.74	100	212.0
8.88	48	118.4	38.85	102	215.6
10.00	50	122.0	40.00	104	219.2
11.10	52	125.6			

[a]For other temperatures, use one of the following:

$$C = (F - 32)\ 5/9 \text{ and } F = (C \times 9/5) + 32$$

or

$$C = 0.555(F - 32) \text{ and } F = 1.8C + 32$$

Table C–2 FRACTIONS, DECIMALS, AND METRIC EQUIVALENTS

Fractions		Decimal Equivalents[a] (in.)	Metric Equivalents[b] (mm)	Metric Equivalents (mm)	Decimal Equivalents (in.)
	1/64	0.0156	0.397	1	0.039
1/32		0.0313	0.794	2	0.079
	3/64	0.0469	1.191	3	0.118
1/16		0.0625	1.588	4	0.157
	5/64	0.0781	1.984	5	0.197
3/32		0.0938	2.381	6	0.236
	7/64	0.1094	2.778	7	0.276
1/8		0.1250	3.175	8	0.315
	9/64	0.1406	3.472	9	0.354
5/32		0.1563	3.969	10	0.394
	11/64	0.1719	4.366	11	0.433
3/16		0.1875	4.763	12	0.472
	13/64	0.2031	5.159	13	0.512
7/32		0.2188	5.556	14	0.551
	15/64	0.2344	5.953	16	0.630
1/4		0.2500	6.350	17	0.669
	17/64	0.2656	6.747	18	0.709
9/32		0.2813	7.144	19	0.748
	19/64	0.2969	7.541	20	0.787
5/16		0.3125	7.938	21	0.827
	21/64	0.3281	8.334	22	0.866
11/32		0.3438	8.732	23	0.906
	23/64	0.3594	9.128	24	0.945
3/8		0.3750	9.525	25	0.984
	25/64	0.3906	9.922	25.4	1.000
13/32		0.4063	10.319		
	27/64	0.4219	10.716		
7/16		0.4375	11.113		
	29/64	0.4531	11.509		
15/32		0.4688	11.906		
	31/64	0.4844	12.303		
1/2		0.5000	12.700		
	33/64	0.5156	13.097		
17/32		0.5313	13.494		
	35/64	0.5469	13.891		
9/16		0.5625	14.288		
	37/64	0.5781	14.684		
19/32		0.5938	15.081		
	39/64	0.6094	15.478		
5/8		0.6250	15.875		
	41/64	0.6406	16.272		

Table C-2 (continued)

Fractions		*Decimal Equivalents*[a] *(in.)*	*Metric Equivalents*[b] *(mm)*
21/32		0.6563	16.669
	43/64	0.6719	17.066
11/16		0.6875	17.463
	45/64	0.7031	17.859
23/32		0.7188	18.256
	47/64	0.7344	18.653
3/4		0.7500	19.050
	49.64	0.7656	19.447
25/32		0.7813	19.844
	51/64	0.7969	20.241
13/16		0.8125	20.638
	53/64	0.8281	21.034
27/32		0.8438	21.431
	55/64	0.8594	21.828
7/8		0.8750	22.225
	57/64	0.8906	22.622
29/32		0.9063	23.019
	59/64	0.9219	23.426
15/16		0.9375	23.813
	61/64	0.9531	24.209
31/32		0.9688	24.606
	63/64	0.9844	25.003

[a]Decimal-to-metric conversion: inches × 25.4 = mm.
[b]Metric-to-decimal conversion: 0.03939 × mm = inches.

Table C-3 BASIC COLOR-CODE SYSTEM

Color	*Significant Figure*	*Decimal Multiplier*	*Tolerance*	*Voltage Rating*	*Characteristic*
Black	0	1	± 20%	—	A
Brown	1	10	± 1%	100	B
Red	2	100	± 2%	200	C
Orange	3	1,000	± 3%	300	D
Yellow	4	10,000	± GMV[a]	400	E
Green	5	100,000	± 5%	500	F
Blue	6	1,000,000	± 6%	600	G
Violet	7	10,000,000	± 12.5%	700	—
Gray	8	0.01	± 30%	800	I
White	9	0.1	± 10%	900	J
Gold	—	0.1	± 5%	1000	—
Silver	—	0.01	± 10%	2000	—
No color	—	—	± 20%	500	—

[a]GMV, guaranteed minimum value.

Table C-4 FILAMENT AND HOOK-UP WIRE CURRENT RATINGS

Hook-up Wire[a]		*Filament Wire*	
Size (AWG)	*Current (A)*	*Size (AWG)*	*Current (A)*
8	30	8	20
10	20	10	12
12	15	12	7
14	10	14	5
16	7	16	3
18	5	18	2
20	4	20	1

[a]The hook-up wire values have been derated to satisfy close-spacing and temperature problems.

Table C–5 COLOR CODE INFORMATION FOR RESISTORS AND CAPACITORS

	Temperature Coefficient		1st digit	2nd digit	3rd digit	Multiplier	Percent of tolerance	Tolerance		Voltage rating	Other characteristic	Temp. °C and characteristic	Percent of tolerance	Other characteristic	Reliability
	EIA	MIL						10pf or less	over 10pf						
	A	A	B	C	D	M	T	F	F	V	O	X	Y	Z	W
Black	0	0	0	0	0	10^0	± 20	± 2	± 20			85–A	± 20	P	
Brown	−33	−30	1	1	1	10^1	± 1	± .1	± 1	100	J	85–E		Q	1.0
Red	−75	−80	2	2	2	10^2	± 2		± 2	200	K			R	0.1
Orange	−150	−150	3	3	3	10^3	± 3		± 2.5	300	L		± 30	S	0.01
Yellow	−220	−220	4	4	4	10^4	± 4			400	M			T	.001
Green	−330	−330	5	5	5	10^5	± 5	± .5	± 5	500	N				
Blue	−470	−470	6	6	6	10^6	± 6			600					
Violet	−750	−750	7	7	7	10^7	± 7			700					
Gray	+150 to−1500	+30	8	8	8	10^8	± 8	± .25		800					
White	+100 to −750	+300	9	9	9	10^9	± 9	± 1	± 10	900					
Gold		+100				10^{-1}	± 5			1000					
Silver						10^{-2}	± 10						± 10		
No color							± 20								

Note: Use this table in conjunction with Figure 211.

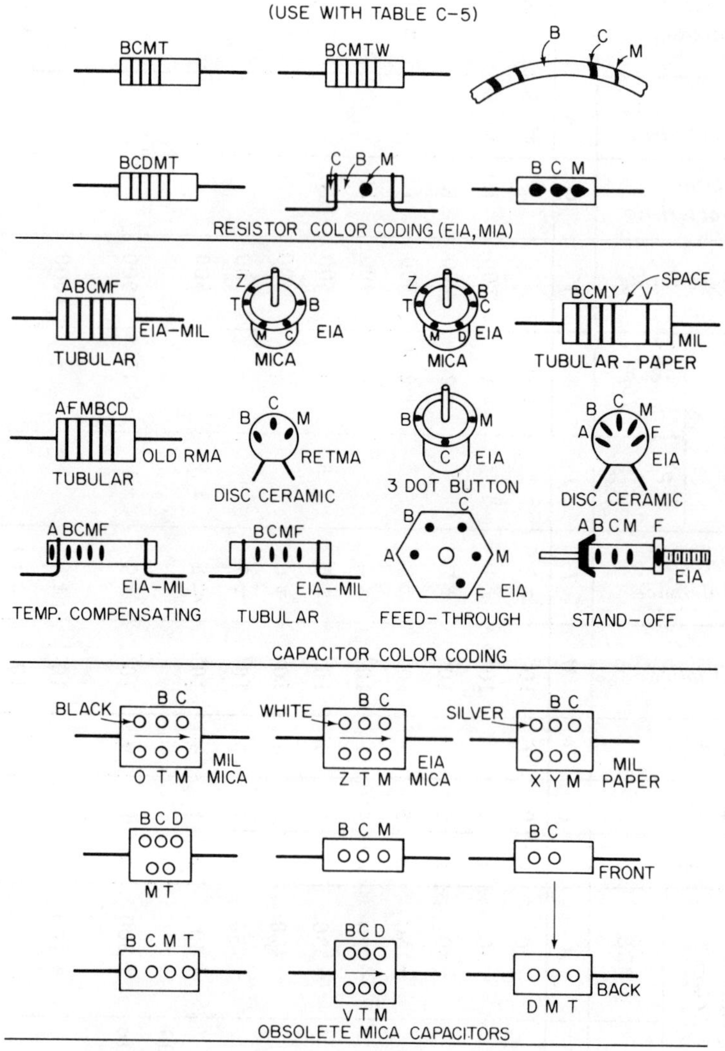

Figure 211 Resistor and capacitor color coding

Table C-6 HOOK-UP WIRE COLOR CODE

Color of Wire	*Circuit Use*
Black	Grounds (grounded elements and returns)
Brown	Heaters and filaments off ground
Red	Power supply (B+), positive to ground
Orange	Screen grids
Yellow	Cathodes, junction transistor *emitter*, or *source* of FET
Green	Control grids or signal leads, junction transistor *base*, or *gate* of FET
Blue	Plates (not tied to B+), junction transistor *collector*, or *drain* of FET
Violet	Supply voltage, negative compared to ground
Gray	AC power and interlock wiring
White	Above- or belowground returns, AVC, bias, and neutrals

Table C-7 RESISTOR VALUES COMMONLY AVAILABLE

5%	*10%*	*20%*
10	10	10
11		
12	12	
13		
15	15	15
16		
18	18	
20		
22	22	22
24		
27	27	
30		
33	33	33
36		
39	39	
43		
47	47	47
51		
56	56	
62		
68	68	68
75		
82	82	
91		
100	100	100

Table C–8 WIRE SIZE AND CURRENT RATINGS

Wire Size (AWG) (B & S)	*Diameter (mils)*	*Circular-Mil Area*	*Maximum Current (A) for Wire Size*
1	289.3	83,694	140
2	257.6	66,357	120
3	229.4	52,624	75
4	204.3	41,738	70
5	181.9	33,087	56
6	162.0	26,244	48
7	144.2	20,794	42
8	128.5	16,512	35
9	114.4	13,088	30
10	101.9	10,384	26
11	90.74	8,234	21
12	80.81	6,530	18
13	72.00	5,174	16
14	64.08	4,106	13
15	57.07	3,257	12
16	50.82	2,583	10
17	45.26	2,048	8
18	40.30	1,624	7
19	35.89	1,288	6
20	32.00	1,024	5
21	28.46	810.0	4
22	25.35	642.6	3
23	11.57	509.4	2.5
24	20.10	404.0	2.0
25	17.90	320.4	1.5
26	15.94	254.1	1.2
27	14.20	201.5	1.0
28	12.64	159.8	0.7
29	11.26	126.8	0.5
30	10.03	100.6	0.4
31	8.928	79.70	0.3
32	7.950	63.20	0.25
33	7.080	50.13	0.20
34	6.305	39.75	0.15
35	5.615	31.53	0.12
36	5.000	25.00	0.10
37	4.453	19.83	0.08
38	3.965	15.72	0.07
39	3.531	12.47	0.05
40	3.145	9.89	0.04

Wire and Sheet-Metal Gauges

In the past many different gauge designations have been used for the description of steel wire, music wire, nonferrous wire, telephone wire, sheet metals, and drill rod. The resulting confusion is lessened by the fact that many of the older gauge designations are now obsolete. In the electronics field the *American Wire Gauge* (known also as the Brown and Sharpe gauge) is the currently accepted standard for nonferrous wire (except telephone wire) and *sheet material.* For iron and steel sheets, use the United States Standard gauge.

Table C–9 WIRE AND SHEET-METAL GAUGES

Gauge Numbers	*American or Brown and Sharpe (Nonferrous Metals)*	*U.S. Standard (Iron and Steel Sheet)*
7/0	0.6513	
6/0	0.5800	0.4687
5/0	0.5165	0.4375
4/0	0.4600	0.4062
3/0	0.4096	0.375
2/0	0.3648	0.3437
1/0	0.3249	0.3125
1	0.2893	0.2812
2	0.2576	0.2656
3	0.2294	0.25
4	0.2043	0.2344
5	0.1819	0.2187
6	0.1620	0.2031
7	0.1442	0.1875
8	0.1285	0.1719
9	0.1144	0.1562
10	0.1019	0.1406
11	0.0907	0.125
12	0.0808	0.1094
13	0.0720	0.0937
14	0.0641	0.0781
15	0.0571	0.0703
16	0.0508	0.0625
17	0.0452	0.05625
18	0.0403	0.05
19	0.0359	0.04375
20	0.032	0.0375
21	0.0285	0.03437

Table C-9 (continued)

Gauge Numbers	*American or Brown and Sharpe (Nonferrous Metals)*	*U.S. Standard (Iron and Steel Sheet)*
22	0.0253	0.03125
23	0.0226	0.02812
24	0.0201	0.025
25	0.0179	0.02187
26	0.0159	0.01875
27	0.0142	0.01718
28	0.0126	0.01562
29	0.0113	0.01406
30	0.01	0.0125

Table C-10 WIRE SIZE AND RESISTANCE

AWG No.	*Diameter (in.)*	*Ohms per 1000 ft*[a]	*AWG No.*	*Diameter (in.)*	*Ohms per 1000 ft*[a]
40	0.00315	1049.0	18	0.04030	6.385
38	0.00397	659.6	16	0.05082	4.016
36	0.00500	414.8	14	0.06408	2.525
34	0.00631	260.9	12	0.08081	1.588
32	0.00795	164.1	10	0.1019	0.999
30	0.01003	103.2	8	0.1285	0.628
28	0.01264	64.9	6	0.1620	0.395
26	0.01594	40.8	4	0.2043	0.249
24	0.02010	25.7	2	0.2576	0.156
22	0.02535	16.1			
20	0.03196	10.15			

[a]Standard annealed copper at 20°C.

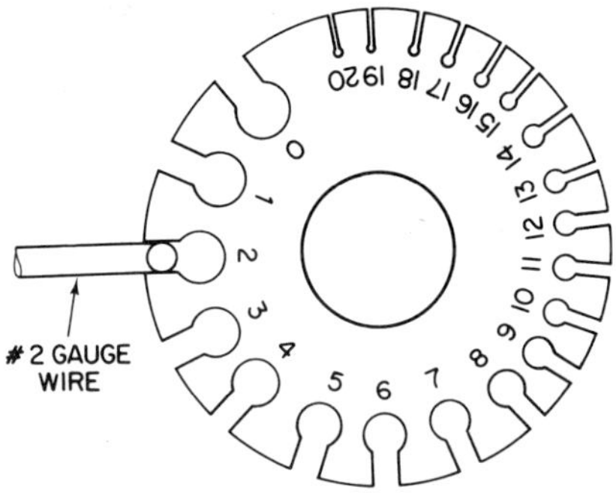

Figure 212 Measuring a #2 wire with a wire gauge tool

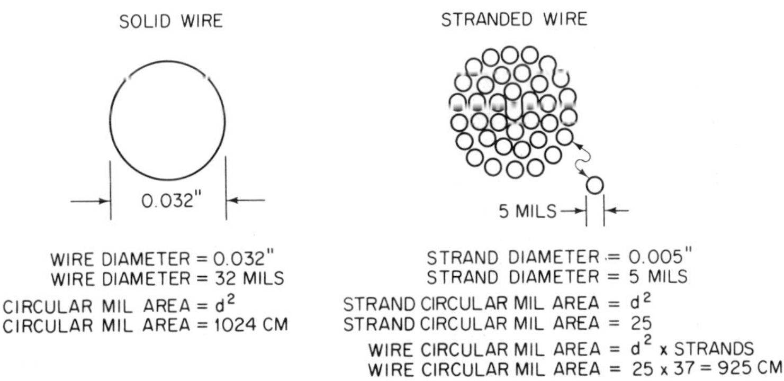

Figure 213 Calculating wire size

Table C-11 TAP-DRILL CHART (American National Form Screw Threads)

Size—Number of Threads per Inch	*Minor Diameter*	*Tap Drill to Give 75% Full Thread*	*Tap-Drill Size*[a]
0-80	0.0438	3/64	0.0469
1-64	0.0527	53	0.0595
72	0.0550	53	0.0595
2-56	0.0628	50	0.0700
64	0.0657	50	0.0700
3-48	0.0719	47	0.0785
56	0.0758	45	0.0820
4-36	0.0764	44	0.0860
40	0.0795	43	0.0890
48	0.0849	42	0.0935
5-40	0.0925	38	0.1015
44	0.0955	37	0.1040
6-32	0.0974	36	0.1065
40	0.1055	33	0.1130
8-32	0.1234	29	0.1360
36	0.1279	29	0.1360
10-24	0.1359	25	0.1495
32	0.1494	21	0.1590
12-24	0.1619	16	0.1770
28	0.1696	14	0.1820
14-20	0.1770	10	0.1935
24	0.1880	7	0.2010
1/4-20	0.1850	7	0.2010
28	0.2036	3	0.2130
5/16-18	0.2403	F	0.2570
24	0.2584	I	0.2720
3/8-16	0.2938	5/16	0.3125
24	0.3209	Q	0.3320
7/16-14	0.3447	U	0.3680
20	0.3726	25/64	0.3906
1/2-13	0.4001	27/64	0.4219
20	0.4351	29/64	0.4531
9/16-12	0.4542	31/64	0.4844
18	0.4903	33/64	0.5156
5/8-11	0.5069	17/32	0.5312
18	0.5528	37/64	0.5781
3/4-10	0.6201	21/32	0.6562
16	0.6688	11/16	0.6875
7/8-9	0.7307	49/64	0.7656
14	0.7822	13/16	0.8125
1-8	0.8376	7/8	0.8750

[a]Tap drill size = major diameter $- \dfrac{1}{\text{threads per inch}}$.

Table C-12 BODY-DRILL CHART (American National Form Screw Threads)

Size–Number of Threads per Inch	*Major Diameter*	*Body Drill*	*Body-Drill Size*
0–80	0.0600	52	0.0635
1–64	0.0730	48	0.0760
72	0.0730	48	0.0760
2–56	0.0860	44	0.0860
64	0.0860	44	0.0860
3–48	0.0990	39	0.0995
56	0.0990	39	0.0995
4–36	0.1120	33	0.1130
40	0.1120	33	0.1130
48	0.1120	33	0.1130
5–40	0.1250	30	0.1285
44	0.1250	30	0.1285
6–32	0.1380	28	0.1380
40	0.1380	28	0.1380
8–32	0.1640	19	0.1660
36	0.1640	19	0.1660
10–24	0.1900	11	0.1910
32	0.1900	11	0.1910
12–24	0.2160	7/32	0.2187
28	0.2160	7/32	0.2187
14–20	0.2400	C	0.2420
24	0.2400	C	0.2420
1/4–20	0.2500	1/4	0.2500
28	0.2500	1/4	0.2500
5/16–18	0.3125	5/16	0.3125
24	0.3125	5/16	0.3125
3/8–16	0.3750	3/8	0.3750
24	0.3750	3/8	0.3750
7/16–14	0.4375	7/16	0.4375
20	0.4375	7/16	0.4375
1/2–13	0.5000	1/2	0.5000
20	0.5000	1/2	0.5000
9/16–12	0.5625	9/16	0.5625
18	0.5625	9/16	0.5625
5/8–11	0.6250	5/8	0.6250
18	0.6250	5/8	0.6250
3/4–10	0.7500	3/4	0.7500
16	0.7500	3/4	0.7500
7/8–9	0.8750	7/8	0.8750
14	0.8750	7/8	0.8750
1–8	1.0000	1	1.0000

Table C–13 LETTER AND NUMBER TWIST DRILLS

Letter Size	*Decimal Size*	*Number Size*	*Decimal Size*	*Number Size*	*Decimal Size*	*Number Size*	*Decimal Size*
Z	0.413	1	0.2280	28	0.1405	55	0.0520
Y	0.404	2	0.2210	29	0.1360	56	0.0465
X	0.397	3	0.2130	30	0.1285	57	0.0430
W	0.386	4	0.2090	31	0.1200	58	0.0420
V	0.377	5	0.2055	32	0.1160	59	0.0410
U	0.368	6	0.2040	33	0.1130	60	0.0400
T	0.358	7	0.2010	34	0.1110	61	0.0390
S	0.348	8	0.1990	35	0.1100	62	0.0380
R	0.339	9	0.1960	36	0.1065	63	0.0370
Q	0.332	10	0.1935	37	0.1040	64	0.0360
P	0.323	11	0.1910	38	0.1015	65	0.0350
O	0.316	12	0.1890	39	0.0995	66	0.0330
N	0.302	13	0.1850	40	0.0980	65	0.0320
M	0.295	14	0.1820	41	0.0960	68	0.0310
L	0.290	15	0.1800	42	0.0935	69	0.02925
K	0.281	16	0.1770	43	0.0890	70	0.0280
J	0.277	17	0.1730	44	0.0860	71	0.0260
I	0.272	18	0.1695	45	0.0820	72	0.0250
H	0.266	19	0.1660	46	0.0810	73	0.0240
G	0.261	20	0.1610	47	0.0785	74	0.0225
F	0.257	21	0.1590	48	0.0760	75	0.0210
E	0.250	22	0.1570	49	0.0730	76	0.0200
D	0.246	23	0.1540	50	0.0700	77	0.0180
C	0.242	24	0.1520	51	0.0670	78	0.0160
B	0.238	25	0.1495	52	0.0635	79	0.0145
A	0.234	26	0.1470	53	0.0595	80	0.0135
		27	0.1440	54	0.0550		

Table C–14 FRACTIONAL TWIST DRILLS

Fractional Size	*Decimal Size*	*Fractional Size*	*Decimal Size*
1/64	0.015625	33/64	0.515625
1/32	0.3125	17/32	0.53125
3/64	0.046875	35/64	0.546875
1/16	0.0625	9/16	0.5625
5/64	0.078125	37/64	0.578125
3/32	0.09375	19/32	0.59375
7/64	0.109375	39/64	0.609375
1/8	0.125	5/8	0.625
9/64	0.140625	41/64	0.640625
5/32	0.15625	21/32	0.65625
11/64	0.171875	43/64	0.671875
3/16	0.1875	11/16	0.6875
13/64	0.203125	45/64	0.703125
7/32	0.21875	23/32	0.71875
15/64	0.234375	47/64	0.734375
1/4	0.25	3/4	0.75
17/64	0.265625	49/64	0.765625
9/32	0.28125	25/32	0.78125
19/64	0.296875	51/64	0.796875
5/16	0.3125	13/16	0.8125
21/64	0.328125	53/64	0.828125
11/32	0.34375	27/32	0.84375
23/64	0.359375	55/64	0.859375
3/8	0.375	7/8	0.875
25/64	0.390625	57/64	0.890625
13/32	0.40625	29/32	0.90625
27/64	0.421875	59/64	0.921875
7/16	0.4375	15/16	0.9375
29/64	0.453125	61/64	0.953125
15/32	0.46875	31/32	0.96875
31/64	0.484375	63/64	0.984375
1/2	0.5	1	1.0

Table C–15 COATED-ABRASIVE GRIT SIZES

	Garnet, Aluminum Oxide, or Silicon Carbide		Emery	Flint
	600,500	(10/0)		
	400	(10/0)		
	360			
Very fine	320	(9/0)		
	280	(8/0)		
	240	(7/0)		
	220	(6/0)		Extra fine
	180	(5/0)	3/0	
Fine	150	(4/0)	2/0	
	120	(3/0)		Fine
			0	
	100	(2/0)		
			½	
Medium	80	(0)	1	Medium
			1½	
	60	(½)		
			2	
	50	(1)		Coarse
Coarse			2½	
	40	(1½)		
	36	(2)		
			3	Extra coarse
	30	(2½)		
	24	(3)		
Very coarse	20	(3½)		
	16	(4)		
	12	(4½)		

Table C–16 CRIMPED SOLDERLESS-CONNECTOR TEST

AWG Wire Size	*Test Current (A)*	*Pull Test (lbs)*	*Maximum Voltage Drop (mV)*	
			Before Pull Test	*After Pull Test*
22	9	8	13	15
20	11	7	12	19
18	16	7	12	38
16	22	7	12	50
14	32	6	11	70
12	41	5	8	110
10	55	5	8	180

Table C–17 DRILL SPEEDS AND LUBRICANTS[a]

Material	*Speed (ft/min)*	*Recommended Cutting Compound*
Aluminum, copper	250	Kerosene, kerosene and lard oil, soluble oil
Brass	200	Dry, soluble oil, kerosene, lard oil
Bronze	200	Dry, mineral oil, soluble oil, lard oil
Cast iron (soft)	120	Dry, air jet
Soft steel	110	Soluble oil, mineral lard oil, sulfurized oil, lard oil
Bakelite	110	Dry
Malleable iron	85	Dry, soda water
Cast iron (hard)	80	Dry, air jet
Tool steel	60	Soluble oil, mineral lard oil, sulfurized oil
Steel forgings	60	Soluble oil, mineral lard oil, sulfurized oil
Monel metal	50	Lard oil, soluble oil
Stainless steel	30	Sulfurized oil, mineral lard oil, soluble oil
Chilled cast iron	20	Dry, black ebony oil
Slate	15	Dry
High-manganese steel	15	Dry

[a]Carbon-steel drills should run at 50% of speeds above; drill-speed formula: rpm = cutting speed ¼ diameter of drill.

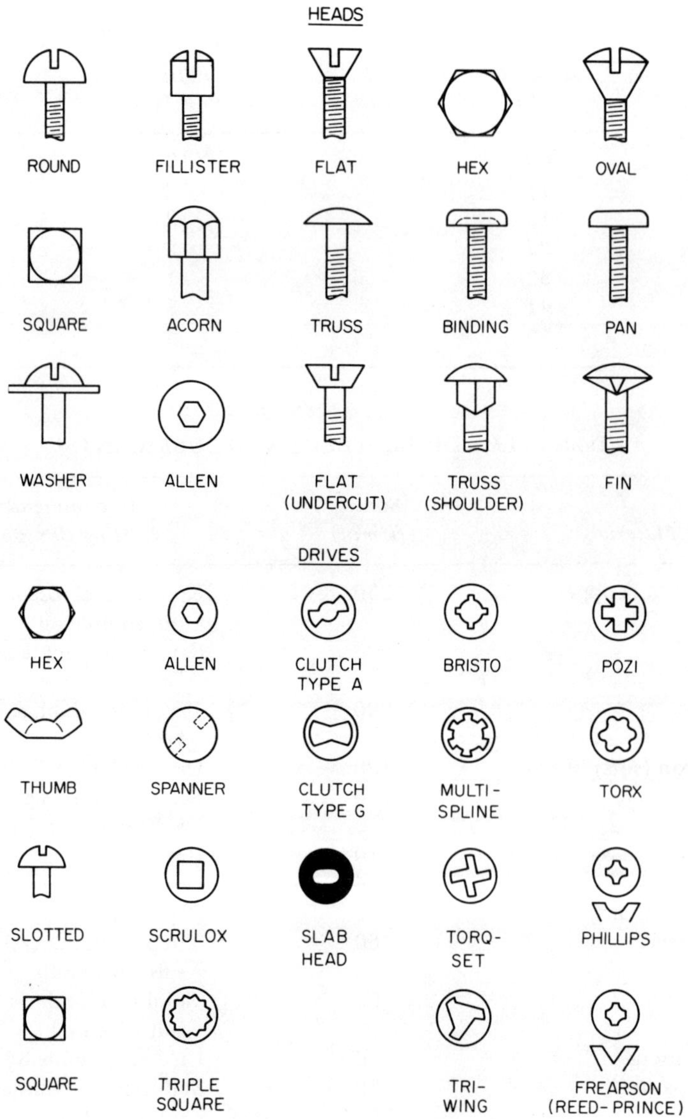

Figure 214 Bolt or screw heads and drives

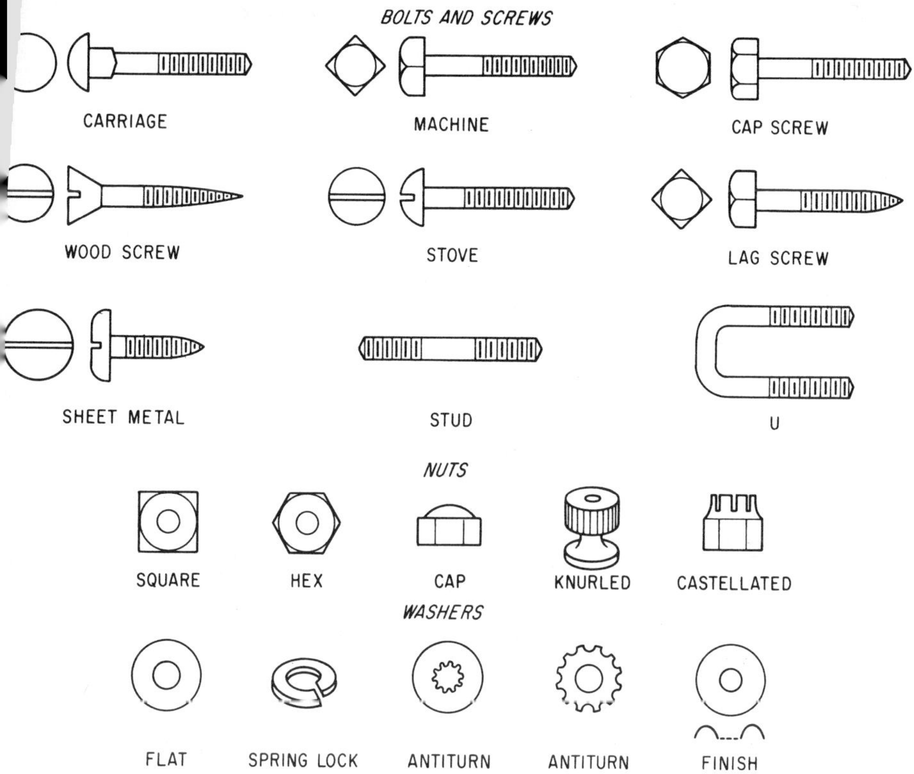

Figure 215 Bolts, nuts, and washers

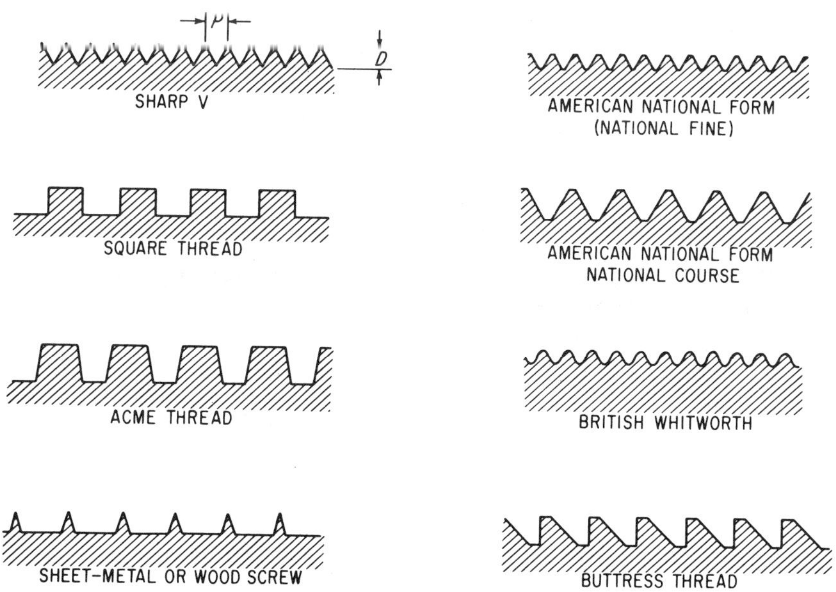

Figure 216 Screw-thread profiles

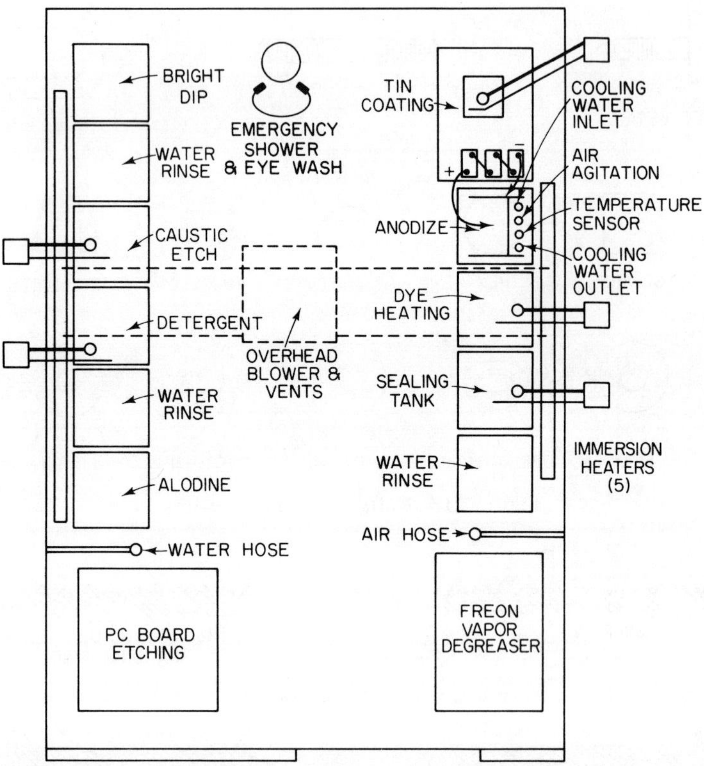

Figure 217 Suggested layout for chemical processing and printed-circuit board fabrication